Moritz Fleischmann

Quantitative Models for Reverse Logistics

Springer

Author

Dr. Moritz Fleischmann
Rotterdam School of Management
Faculteit Bedrijfskunde
Erasmus University Rotterdam
Burgemeester Oudlaan 50
3000 DR Rotterdam
The Netherlands

Cataloging-in-Publication data applied for

Die Deutsche Bibliothek - CIP-Einheitsaufnahme

Fleischmann, Moritz:
Quantitative models for reverse logistics / Moritz Fleischmann. -
Berlin ; Heidelberg ; New York ; Barcelona ; Hong Kong ; London ;
Milan ; Paris ; Singapore ; Tokyo : Springer, 2001
(Lecture notes in economics and mathematical systems ; 501)
ISBN 3-540-41711-7

ISSN 0075-8450
ISBN 3-540-41711-7 Springer-Verlag Berlin Heidelberg New York

Springer-Verlag Berlin Heidelberg New York
a member of BertelsmannSpringer Science+Business Media GmbH

http://www.springer.de

Printed in Germany

Typesetting: Camera ready by author
Printed on acid-free paper SPIN: 10798883 55/3142/du 5 4 3 2 1 0

Acknowledgement

The material presented in this monograph is a result of my work as a PhD student at the Rotterdam School of Management / Faculteit Bedrijfskunde at the Erasmus University Rotterdam in the period of 1996–2000. One of the pleasant things about completing a project like this is to have an opportunity to thank those who have been contributing to its coming into life. I am indebted to many individuals for their support, inspiration, and encouragement, which have made the past four years such a rich experience to me.

First of all, I would like to thank my promotors Jo van Nunen and Rommert Dekker for their confidence and support. Jo has been a great boss to whom I am grateful for creating an excellent working environment and for his continuous spurs to diverge from traditional perspectives. Rommert's enthusiasm and commitment have been an example which I am determined to remember in my future path in research. In the same breath, I would like to thank Roelof Kuik for all the time and patience he put into accompanying and advising me in my research. Countless 'blazing through'–sessions ranging from Markov decision theory to Dutch politics have been a main flavouring of my time as a PhD student.

Furthermore, I am grateful to Luk Van Wassenhove for his valuable advice throughout the past four years and for giving me the opportunity to spend a research visit at INSEAD. I am indebted to Marc Salomon for inviting me to Rotterdam and for introducing me to the research area of Reverse Logistics and product recovery management.

A joint study with IBM has been an essential element of my research project. I would like to thank all colleagues at the International Maintenance Parts Logistics department in Amsterdam for their support and their pleasant co–operation. In particular, I am grateful to Doron Limor for giving me the opportunity to investigate Reverse Logistics practice at IBM and for his open–minded manner of co–operation.

In the same vein, I thank all colleagues I have had the privilege to work together with. In particular, exchange with members of the REVLOG network has been a great opportunity. Based on this experience, I have perceived co–operation and the exchange of ideas as some of the most attractive facets of scientific research.

It goes without saying that a sound 'home base' is invaluable in providing a pleasant and fruitful working environment. Therefore, I would like to thank all colleagues at the Department of Decision and Information Sciences, my fellow PhD candidates at the Rotterdam School of Management / Faculteit Bedrijfskunde, and our friends in Rotterdam and elsewhere in The Netherlands. It is the warm welcome and the many personal relations that have made the past four years such a pleasant experience and that have rendered the notion 'abroad' almost insignificant.

At the same time, I am grateful to my family and friends in Germany for their lasting interest and friendship even across the longer geographical distance. In particular, I thank my parents for their continuous care and support that I could always rely on.

Most of all, however, I thank my wife Birgit, for her love and confidence and for making 'Rotterdam' our common endeavour from the very beginning on. I dedicate this book to her.

Rotterdam, December 2000 *Moritz Fleischmann*

to Birgit

Contents

Part I

Reverse Logistics: An Introduction

1. Introduction

In a world of finite resources and disposal capacities, recovery of used products and materials is key to supporting a growing population at an increasing level of consumption. As waste reduction is becoming a major concern in industrialised countries a concept of material cycles is gradually replacing a 'one way' perception of economy. Increasingly, customers expect companies to minimise the environmental impact of their products and processes. Moreover, legislation extending producers' responsibility has become an important element of public environmental policy. Several countries, in particular in Europe, have introduced environmental legislation charging manufacturers with responsibility for the whole lifecycle of their products. Take–back and recovery obligations have been enacted or are underway for a number of product categories including electronic equipment in the European Union and in Japan, cars in the European Union and in Taiwan, and packaging material in Germany. At the same time, companies are recognising opportunities for combining environmental stewardship with plain financial benefits, brought about by production cost savings and access to new market segments. In this vein, the past decade has seen an explosive growth of product recovery activities both in scope and scale. Consider the following illustrative examples:

- Major manufacturers of copy machines, such as Xerox, Canon, and Océ all are putting substantial effort into remanufacturing used equipment. Xerox reports on annual savings of several hundred million dollars due to remanufacturing and reuse of equipment and parts, while at the same time diverting more than fifty thousand tonnes of material from the waste stream. Ninety percent of today's equipment is claimed to be remanufacturable. In addition, close to seventy percent of Xerox's copy and print cartridges are retrieved for reuse and recycling in Europe and in the USA (Xerox, 1998). Canon has been operating two remanufacturing factories for used copy machines in Virginia (USA) and in the UK since 1993 and is currently exploring comprehensive recycling systems for all copier parts. Toner cartridges have been collected for reuse since 1990. By 1997 three factories in Europe and Asia had recovered more than 20 million cartridges (Canon, 1998). Similar initiatives have been taken by Océ (Océ, 1998).
- In the chemical industry several companies have recently engaged in the recycling of used carpeting. In 1999 DSM Chemicals and Allied Signals

jointly opened a large–scale recycling plant in Georgia (USA) to recover nylon raw material from carpet waste. The facility, involving investment costs of about USD 80 million, has an annual processing capacity of 90,000 tonnes and will process nylon carpet waste collected from 75 metropolitan areas in the USA. Feasibility of a similar system in Europe on an even larger scale has been investigated in a recent study in co–operation with the European carpet industry (DSM, 1999). Dupont operates a similar system in the USA. A recycling plant in Tennessee, which was opened in 1995, is dedicated to recovering nylon material from commercial carpeting for reuse in different applications, including new carpet fibres and car parts. The facility has the capacity to produce up to fifteen hundred tonnes of nylon material annually (Dupont, 1999).

- Yet another example of product recovery concerns single–use cameras. In response to heavy environmental criticism Kodak started in 1990 to take back, reuse, and recycle its single–use cameras, which had originally been designed as disposables. Collection volumes grew from 0.9 million cameras in 1990 to 61 million cameras in 1998. Today, three recycling facilities are operated world–wide and up to 86 percent of the cameras' parts are reused in manufacturing new cameras (Kodak, 1999).
- The growing significance of product recovery is also becoming apparent on a macro–economic level. For example, glass recycling rates, as percentage of consumption, grew from 5% in 1980 to more than 26% in 1997 in the USA, from 20% to 52% in France, and even from 23% to 79% in Germany. In the same period, the rate of paper recycling jumped from 20% to 40% in the USA, from 30% to 41% in France and from 34% to 70% in Germany (OECD, 1999). In the Netherlands, almost 70% of all industrial waste was recovered in 1996 rising from 36% in 1992 (CBS, 1999).

For a long time, research concerning product recovery has focussed on engineering and, to a lesser extent, marketing issues. It is only in the past decade that the need to investigate the logistics aspects of reuse and recycling has been generally recognised. From a logistics perspective, product recovery initiates additional goods flows from users to producers. The management of these flows, opposite to the conventional supply chain flows, is addressed in the recently evolved field of 'Reverse Logistics'. However, recovery of used products is not the only reason for goods flows 'going the other way round'. Returns of unsold merchandise from retailers to manufacturers is another example of growing importance. Shifting channel power increasingly is forcing manufacturers to take back and refund overstocks in the supply chain. Furthermore, returns of defective products or parts for repair is a very old example of 'reverse' goods flows. In view of these diverse examples we address the delineation of 'Reverse Logistics' in more detail in the next section. Subsequently, we formulate our research objectives in Section 1.2 and discuss our methodology. Section 1.3 then provides an outline of the book.

1.1 Scope and Definition of Reverse Logistics

Given the short history of research on the logistics aspects of product recovery it may not come as a surprise to note a lack of generally accepted terminology. In particular, we find the term 'Reverse Logistics' used in different meanings. To illustrate the different connotations we consider the definitions formulated in four prominent publications.

In one of the first publications referring to this term, a White Paper published by the Council of Logistics Management (CLM), Reverse Logistics is introduced as

> " [...] the term often used to refer to the role of logistics in recycling, waste disposal, and management of hazardous materials; a broader perspective includes all issues relating to logistics activities carried out in source reduction, recycling, substitution, reuse of materials and disposal." (Stock, 1992)

A similar characterisation is given by Kopicki et al. (1993). In yet another early paper Pohlen and Farris (1992) define Reverse Logistics as

> " [...] the movement of goods from a consumer towards a producer in a channel of distribution." (Pohlen and Farris, 1992)

More recently, Rogers and Tibben–Lembke (1999) have alluded to the CLM's definition of logistics by defining Reverse Logistics as

> " [...] the process of planning, implementing, and controlling the efficient, cost–effective flow of raw materials, in–process inventory, finished goods, and related information from the point of consumption to the point of origin for the purpose of recapturing value or proper disposal." (Rogers and Tibben-Lembke, 1999)

Note that each of these definitions refers to a different criterion for delineating Reverse Logistics. Stock and Kopicky et al. emphasize the element of waste reduction and place Reverse Logistics in the context of environmental management. In contrast, Pohlen and Farris refer to the 'direction' of a goods flow, relative to the supply chain positions of the sender and the receiver. Finally, Rogers and Tibben–Lembke look at the management of goods flows that lead to a closed loop in the supply chain.

The domains of the three definitions overlap to a substantial extent. In particular, all three include flows of used goods returned to the original manufacturer. However, it should be noted that the definitions are not identical and that none of them implies the others. For example, the first definition, in contrast with the latter, appears not to include flows of new products returned for commercial reasons. On the other hand, the second definition does not include product returns to other parties than producers, such as packaging returns to retailers, whereas the other two do. Finally, the third definition excludes flows of used products to specialised recovery companies, such as independent remanufacturers, whereas the other two both include them.

In view of these disparate definitions we feel the need to give our own interpretation of the term 'Reverse Logistics' as we use it throughout this manuscript. To this end, we strive for a delineation that encompasses objects of similar characteristics, that allows us to study logistics decision problems, and that is in line with current practice. In this light, we characterise Reverse Logistics by the following three aspects.

- First of all, we see Reverse Logistics as an element of the growing diversity of logistics systems. Traditionally, supply chains have been perceived as unidirectional structures with a well defined hierarchy. As noted by Ganeshan et al. (1998) "...the most common definition [of a supply chain] is a system of suppliers, manufacturers, distributors, retailers, and customers where materials flow downstream from suppliers to customers and information flows in both directions." However, we currently see logistics systems developing into general networks of organisations that cannot be ordered in such a way that all materials flow 'downstream'. In particular, the examples in the previous section illustrate the growing importance of material flows opposite to the traditional supply chain direction. While these 'reverse' flows should not be segregated from the overall picture they deserve focused attention, and hence a distinct 'name', as they bring about novel business situations and management issues. As experience accumulates one may expect the distinction between 'forward' and 'reverse' flows to fade away, making room for a holistic supply chain perception.
- Second, Reverse Logistics is concerned with secondary goods flows in the sense that it refers to products of which an original use has been completed or has become impossible. Hence, Reverse Logistics deals with the derivatives of some previous use, which was either planned or actually realised. The objective is to maximise economic value given the resulting products. This may include disposal or some form of recovery.
- Third, the term Reverse Logistics very much expresses the perspective of the receiving party. Hence, Reverse Logistics is a special form of inbound logistics.

We summarise these characteristics in the following definition, which is used as a point of reference in the remainder of the book.

Reverse Logistics is the process of planning, implementing, and controlling the efficient, effective inbound flow and storage of secondary goods and related information opposite to the traditional supply chain direction for the purpose of recovering value or proper disposal.

We illustrate the scope of this Reverse Logistics concept in detail in Chapter 3. To conclude this section let us compare our characterisation with the previous definitions above. Like Stock and Kopicki et al. we refer to logistics management in the context of product recovery and disposal. Note, however, that we use a fairly broad concept of secondary products. In particular, we include unused, and in this sense, new but obsolete products (e.g. overstocks). It is not quite clear whether this category is covered by the former definition. On the other hand, we impose additional supply chain conditions to delineate Reverse Logistics. For example, municipal waste collection is not included as it does not concern flows opposite to the traditional supply chain direction. With Pohlen and Farris we share the criterion of the flow 'direction'. However, rather than restricting Reverse Logistics to flows between consumers and producers we include any 'upstream' flow. Similarly, we generalise Rogers and Tibben-Lembke's definition by including flows other than returning to the 'point of origin', such as product recovery by a competitor or in an alternative supply chain. Finally, we have explicitly added the aspect of inbound logistics, which is not mentioned in any of the other definitions, but which appears to be in line with how the term Reverse Logistics is currently used.

1.2 Research Goals and Methodology

This monograph is concerned with decision making in Reverse Logistics. More specifically, we consider the planning, implementation, and control tasks as delineated in the previous section. In traditional 'forward' logistics, quantitative models have proved a powerful tool for supporting these types of decisions and, more generally, for understanding the underlying systems. For many decision problems standard Operational Research models such as facility location models, routing and scheduling models, or stochastic inventory models have been developed that are widely recognised. Given the short history of the field, a similar set of standard models in a Reverse Logistics context has not yet been established. In particular, the robustness and flexibility of traditional models for coping with Reverse Logistics issues is not yet clear. Although the number of individual contributions has been growing quickly in recent years a comprehensive framework is still lacking. Therefore, the time appears to be right for a systematic analysis of decision making in a Reverse Logistics context. It is the goal of this monograph to contribute to this overall view. In this vein, the thext aims at:

- enhancing the body of well understood quantitative models
- generalising observations from individual Reverse Logistics cases
- contributing to a better understanding of Reverse Logistics issues
- contributing to improved decision making in Reverse Logistics.

Quantitative research in traditional logistics largely focuses on applications, on the one hand, and on efficient solution algorithms, on the other hand; whereas the underlying models are, in general, well established. In Reverse

Logistics the models themselves are a point of debate. They should be well motivated before turning to advanced solution techniques. Therefore, modelling is where we put the focus of our contribution.

To this end, we first investigate which logistics issues arise in the management of 'reverse' goods flows. Given the emerging nature of the field we focus on logistics core functions, namely managing the flow and storage of goods in order to bridge the gap between supply and demand both in space and time. In other words, we pay attention to distribution and inventory management issues in the first place. Other logistics aspects, such as related information management issues, are considered in less detail as we move along.

Having identified relevant Reverse Logistics decision problems we analyse their specific characteristics. In particular, we compare Reverse Logistics settings with traditional 'forward' logistics contexts. While some of these issues have been addressed before, the rapid development of Reverse Logistics in recent years merits a refined up to date picture. On this basis we can then go a step further towards a systematic quantitative analysis. To this end, we investigate how the specific characteristics of Reverse Logistics issues can be translated into appropriate quantitative models. Finally, this allows us to capture the impact of 'reverse' goods flows on logistics systems quantitatively. To summarise, we organise our research around the following questions:

- Which logistics issues arise in the management of 'reverse' goods flows?
- What are the differences between Reverse Logistics and traditional 'forward' logistics and what is the role of Reverse Logistics in an overall logistics concept?
- How can the characteristics of Reverse Logistics appropriately be captured in quantitative models that support decision making?

As a basis for addressing these questions we consider a set of case studies. Recent literature provides insights from a number of examples of Reverse Logistics practice. In addition, we supply first hand experience from a study on Reverse Logistics at IBM. Bringing these cases together we identify common characteristics of Reverse Logistics issues and contrast them with more traditional logistics contexts. Based on this comparison we investigate the proficiency of standard OR models for addressing Reverse Logistics issues. To this end, we summarise specific models proposed in literature and develop new models and extensions as required. In analysing these models we highlight the impact of 'reverse' goods flows on the tradeoffs ruling logistics decision making.

1.3 Outline of this Monograph

The material in this monograph is organised in four main parts. Part I provides a general introduction to Reverse Logistics and lays out a structural framework for our analysis. Parts II and III form the core of this text and concern distribution management and inventory management in a Reverse Logistics context, respectively. For both areas we provide a survey of recent case studies and complement it with a quantitative analysis. Part IV condenses our results and illustrates their application in a business case. More specifically, the content of the individual chapters is as follows:—

Part I
Following these introductory sections *Chapter 2* illustrates the facets of Reverse Logistics in the example of IBM. The role of different kinds of 'reverse' goods flows is indicated and corresponding logistics issues are discussed. The material in this chapter is based on a study in co–operation with IBM's service parts organisation in Amsterdam, The Netherlands.

Chapter 3 structures the field of Reverse Logistics. To this end, we first consider a number of context variables, namely drivers for product returns, dispositioning options, and actors. Subsequently, we indicate different categories of Reverse Logistics flows and discuss their characteristics. This layout serves as a point of reference in the remainder of the book. In addition, literature is surveyed that complements the field of our investigation. In particular, attention is paid to insights from related marketing and production management studies. Part of this material is taken from (Fleischmann et al., 1997a).

Part II
Chapter 4 deals with distribution management issues in Reverse Logistics. Focus is on physical logistics network design. The chapter addresses the question whether Reverse Logistics leads to significantly different network characteristics. To this end, a set of ten recent case studies in literature is reviewed. We identify commonalities and compare them with logistics networks in a more traditional context. Moreover, a network classification is proposed based on different forms of product recovery. In addition, related vehicle routing issues are briefly discussed. The results of this chapter have appeared in adapted form in (Fleischmann et al., 1999) and (Bloemhof–Ruwaard et al., 1998).

Chapter 5 reconsiders the above results from an Operational Research perspective and addresses the question how to capture the identified Reverse Logistics characteristics in quantitative network design models. In most of the aforementioned case studies a corresponding facility location model has been proposed. We discuss particular model properties and put them in relation with the observed network characteristics. In addition, we propose a general modelling framework encompassing different Reverse Logistics contexts. Subsequently, this model is used to investigate the robustness of logistics network structures with respect to incorporating 'reverse' goods flows. The main results of this chapter have recently been discussed in (Fleischmann et al., 2000).

Part III
Analogous with the above approach, *Chapter 6* considers inventory manage-

ment in a Reverse Logistics context. The goal is to identify recurrent issues. We again start from a set of cases studies of which we identify common characteristics. In addition, we review corresponding quantitative models in literature.

Chapter 7 is concerned with the impact of exogenous inbound goods flows on inventory control. Specifically, it analyses the impact of recoverable product returns on appropriate replenishment decisions and on logistics costs. To this end, we formulate a generalised stochastic inventory control model. By using general Markov theory an optimal replenishment policy is derived. Moreover, it is shown that the extended model can be transformed into an equivalent traditional model. This allows for an easy computation of optimal control parameters and minimal system costs. In addition, we address the potential benefit of improved information on future product returns. The material in this chapter refers to (Fleischmann et al., 1997b; Fleischmann and Kuik, 1998; Van der Laan et al., 1998; Heisig and Fleischmann, 1999).

Chapter 8 addresses some modelling extensions. In particular, attention is paid to the issue of multiple supply alternatives, which appears to be typical of many Reverse Logistics situations. It is shown that limited resource availability can lead to counterintuitive effects, namely a loss of monotonicity of the optimal order policy.

Part IV

Chapter 9 applies insights from our analysis to a real–life Reverse Logistics problem at IBM. Specifically, we address the issue of how to incorporate returns of used equipment as a supply source into the spare parts management. The major challenge concerns the tradeoff between procurement cost savings, on the one hand, and uncertain return availability, on the other hand. We design a simulation model for comparing alternative logistics concepts and derive an appropriate solution considering logistics costs and service. The chapter is again based on the aforementioned co–operation with IBM.

Finally, *Chapter 10* summarises our findings and discusses directions for future research.

2. Reverse Logistics at IBM: An Illustrative Case

To highlight the importance of Reverse Logistics in today's business environments and to illustrate emerging issues we discuss an exemplary case in some detail. For this purpose, this chapter considers the Reverse Logistics activities of IBM, one of the major players in the electronics business. At the same time, this sector is one of the most prominent in the recent Reverse Logistics developments. High market volumes, short product life–cycles, and technical feasibility of electronic component reuse due to the absence of 'wear and tear', in contrast to mechanical components, result in a vast product recovery potential. Moreover, disposal of electronic equipment is increasingly being restricted in many countries (see e.g. VROM, 2000).

Unless indicated otherwise the material discussed in this chapter is based on a study in co–operation with IBM's service parts logistics division in Amsterdam, The Netherlands. Details of this study are presented in Chapter 9. For the moment, we restrict ourselves to giving a global picture of IBM's Reverse Logistics activities.

IBM's business activities as a leading manufacturer of IT equipment and services involve several groups of 'reverse' goods flows. The total annual volume amounts to several ten thousand tonnes world–wide. Recall from Section 1.1 that we have delineated Reverse Logistics as concerning inbound flows of secondary goods from downstream supply chain parties. In the case of IBM this encompasses the following categories (see also Schut and Germans, 1997; Dijkhuizen, 1997):

- used machines
  - lease returns
  - trade–in offers
  - environmental take–back
- unused machines
  - retailer stock rotation
  - cancelled orders
- rotable spare parts

The first distinction can be made between returns of entire machines on the one hand and of spare parts on the other. The former can be further divided into used and unused equipment. More specifically, the different groups can be summarised as follows.

Used product returns stem from several sources, mainly in the business market. First of all, the most traditional category concerns returns of leased equipment. In this case, machines are returned to IBM unless customers eventually buy them at the end of the lease period that typically amounts to about three years. Second, IBM may offer to trade in used machines from customers buying new ones. Commercial considerations and the recovery of valuable resources are the major drivers for this initiative. Moreover, in this way knowledge intensive components can be prevented from leaking to broker markets or competitors. Third, IBM has established product take back programs in selected countries, including the USA, South Africa, and most countries in Western Europe, and offers to take back any used equipment customers want to dispose of for free or in return for a small fee. In addition to the above drivers, a 'green' company profile and compliance with current or expected environmental regulation play a major role in this context. The latter concerns both the business and the consumer market. For example, the 'White and Brown Goods Act' in the Netherlands obliges manufacturers and importers of electronic appliances to take back their products after use and recover certain minimum percentages (VROM, 2000). Similar legislation applies in Norway and is currently discussed in several other European countries as well as on an EU–level. Similar directions are also taken in Eastern Asia, including Japan and Taiwan.

On a much smaller scale, IBM also faces *returns of unused machines.* As part of its marketing strategy IBM grants retailers the right, under specific conditions, to return a certain amount of unsold stock against refunding. In other words, IBM covers a part of the retailer's market risk. Similarly, a customer may be allowed to cancel an order until a certain point in the sales process, possibly after the shipment has taken place. Both of these cases again primarily concern the business market.

Finally, another important class of 'reverse' goods flows at IBM concerns *rotable spare parts*, which have been key to IBM's service concept for a long time. To support its service activities IBM maintains stocks of some 100,000 different spare parts. Defective parts replaced in a customer's machine are sent back for repair and can then be used as spare parts again. In this way, keeping parts in a closed loop as much as possible substantially reduces procurement costs. In addition to defective parts, good parts that were needed for diagnostic reasons are also returned.

Recognising the growing strategic impact of Reverse Logistics flows IBM has recently set up a new business unit Global Asset Recovery Services (GARS) that is responsible for managing all goods return flows world–wide. In particular, assigning returned goods to appropriate reuse options, i.e. dispositioning, is an important task of GARS. By centralising these activities in one organisation IBM opts for an active return management that systematically exploits the resource potential of 'reverse' goods flows. In order to recover a maximum of value from the various sorts of returned equipment

IBM considers a hierarchy of reuse options on a product, part, and material level. In this way, goods return flows account for a total annual financial benefit of several hundred million US$. In the sequel we discuss the Reverse Logistics channels per goods category and highlight salient issues. Focus is on the geographic area of Europe, the Middle East, and Africa (EMEA). For America and the Asia and Pacific region similar observations hold.

Used machines
The Reverse Logistics channel for used machines is displayed in Figure 2.1. Machines from business customers are, in general, returned via one of the national distribution centres (DC). Subsequently, several recovery options are considered. If a used machine is deemed remarketable it is assigned to refurbishment. After testing, replacing worn-out or outdated modules, repair, and cleaning it may then be re–sold, possibly via internet. Refurbishment is relevant mainly for lease returns. The selection is made on the basis of the machine type by means of a regularly updated 'keep–list'. Presently, IBM runs two refurbishment facilities in Europe for different product ranges, namely in Montpellier, France and in Niederroden, Germany.

If refurbishment is not viable a used machine may be dismantled in order to recover valuable parts. Parts are tested and repaired if necessary and may then enter the spare parts circuit (discussed in more detail below). If not required internally, some of the generic parts may also be sold to external parties, such as brokers. Used machines are a valuable source for spare parts since the costs for dismantling and subsequent processing are significantly lower than for buying new parts. Moreover, dismantling represents an opportunity to avoid maintaining production capacity for spare parts only, which is relevant since the service period for a given machine type is typically much longer than its period of production. Dismantling is currently carried out at three European IBM locations that process equipment from several EMEA countries, namely in Amsterdam, The Netherlands, in Mainz, Germany, and in Montpellier, France.

The remainder of the used machines is transferred to recycling subcontractors, such as MIREC in the Netherlands, to recover secondary raw materials. On a global scale, only a remainder of some 5% is landfilled or incinerated (see also IBM, 1998).

Used machines returned from the consumer market follow a different road. Since individually collecting this equipment, which usually has a low market value, tends to be inefficient IBM supports branch-wide approaches for this market sector. For example, to comply with recent product take-back legislation in The Netherlands IBM participates in a system organised by ICT, the Dutch association of information and communication technology producers. In this case, used machines from different manufacturers are collected by the municipalities from where they are shipped to recycling companies subcontracted by ICT. Costs for transportation and recycling are shared by the ICT members, proportional to the volume share of their brands in the return flow.

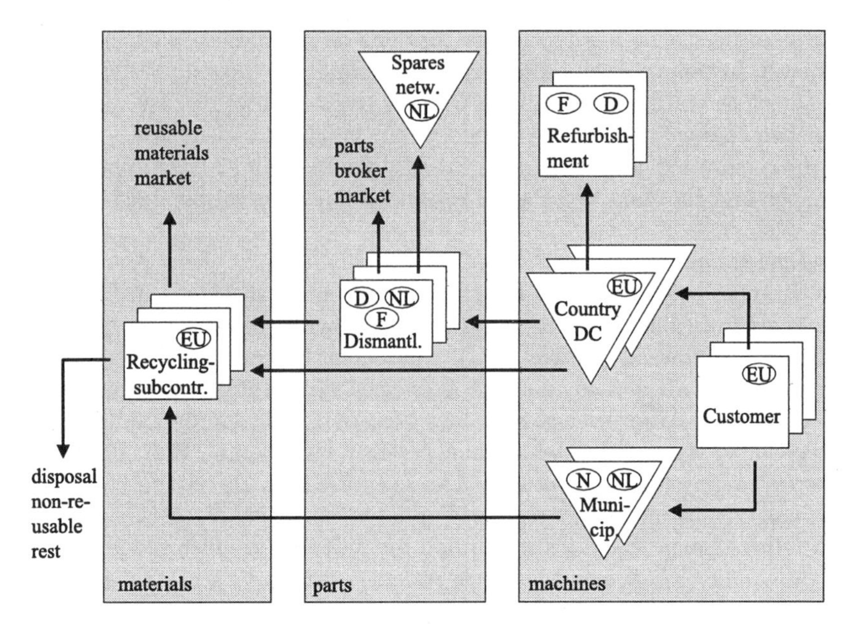

Fig. 2.1. IBM reverse channel for used machines

The current locations of the above IBM recovery facilities are largely historically motivated. Re–designing the corresponding logistics network is considered in the context of the global returns management. In the solution a number of identical return centres should process all used machines from the European market. From this centralisation one may expect a substantial increase in the efficiency and technical viability of the recovery activities. However, an international logistics network design faces legislative difficulties in this context. Both in Europe and in the USA transporting waste across borders is, in many cases, not allowed. Therefore, one needs to determine whether used computer equipment is to be considered as waste or rather as a recoverable resource. It appears that present legislation has been overtaken by some of the rapid advances in product recovery.

In addition, the dispositioning strategy is an important issue. The current fixed hierarchy of options, namely refurbishment before dismantling before recycling may not always be economically optimal. For example, benefits from parts dismantling may, in certain cases, outweigh those from refurbishment, especially since opportunities for actually reselling a refurbished machine are uncertain. Addressing this issue more systematically raises the question of how to value returned goods. Leased equipment has, in principle, a meaningful bookvalue. However, other returned used products are basically obtained 'for free' and do not have a well determined 'market value'. This does not only lead to difficulties in accounting and tax issues but also in financial controlling. Therefore, appropriate ways need to be found to assign costs and benefits of returned goods to activities and organisations. Determining inventory costs is one of the related issues in the logistics context.

Unused machines
Unused equipment is, in general, also returned via the national warehouses. Given the technically 'new' condition of these products one seeks to resell them, in the first place. However, short product life cycles and hence fast depreciation render this option highly time-critical. Therefore, much effort is put into finding alternative markets as fast as possible. Alternatively, returning unused equipment to the manufacturing location is considered. Machines may be disassembled and serve again as input to production processes. Finally, returned unused machines for which no other opportunity is found join the stream of used equipment. Parts dismantling or recycling may then contribute to recovering at least some fraction of the original product value. In the case of unused equipment, clear financial responsibilities prove particularly important in order to maximise the overall result.

Spare parts
The IBM spare parts network for the EMEA region encompasses a hierarchy of stock locations that is fed via a central buffer in Amsterdam, The Netherlands. As discussed above, parts are kept in a closed loop as much as possible. Therefore, defective parts from a customer's machine are returned by an IBM service engineer into the network to be stocked as 'available for repair' in national warehouses. Upon requirement they are then sent to a parts specific central repair location, which may be IBM–owned or an external party. Repaired parts are added to the regular stock again.

At present, the parts return flow largely follows the 'forward' network structure. In order to speed up the return process, and hence to achieve earlier parts availability, a more dedicated reverse channel design is considered. In particular, some levels of the 'forward' network may be bypassed for more direct flows.

As discussed above, dismantling used machines may serve as an alternative source for spare parts. However, efficiently exploiting this source faces a number of difficulties. In particular, uncertainty is a major issue. To a large extent, the return flow of used machines cannot be controlled and is hard to predict. Moreover, even when a machine is available it is not always clear which components it exactly contains, due to intermediate reconfiguration or changes made by the customer. Therefore, dismantling is perceived as a cheap yet uncertain supply source. Furthermore, quality is a major issue in this context. IBM is careful not to corrupt its quality standards by introducing used equipment into its spare parts circuit. In principle, a used machine that is traded in by a customer is not defective. However, certifying its quality may require expensive inspection and testing. Hence, there is a tradeoff between the cost of the dismantling channel and the quality guarantee. We return in detail to the issues around managing dismantling in Chapter 9.

3. Structuring the Field

The objective of this chapter is to lay out a structure of Reverse Logistics serving as a point of reference in the remainder of our investigation. Section 3.1 discusses major dimensions of the Reverse Logistics context, namely drivers, actors, dispositioning options, and cycle times. Based on these aspects, Section 3.2 characterises different categories of Reverse Logistics flows. Section 3.3 concludes the chapter with a literature review complementing the field of our investigation.

3.1 Dimensions of the Reverse Logistics Context

The previous chapters have made clear that Reverse Logistics flows are diverse. Examples range from reusable packages to disposed computer equipment and from returns of unsold merchandise to rotable spare parts. One may think of many criteria for a classification. In the sequel we discuss drivers, dispositioning options, actors, and cycle times as they appear particularly important from a logistics perspective.

Reverse Logistics drivers
Economic, marketing, and legislative motives are commonly cited as reasons for companies to engage in Reverse Logistics. We add asset protection to this list and briefly discuss each of these drivers below.

First of all, 'reverse' inbound flows may be *economically* attractive since used or returned products represent cheap resources from which value may be recovered. Recovery is often cheaper than building or buying new products or 'virgin' materials. In view of low raw material prices, economic attractiveness often relies on the recovery of added manufacturing value rather than on mere material recovery. However, there may be exceptions, such as precious metal recycling.

Second, *marketing* triggers refer to the role of Reverse Logistics in improving a company's market position. On the one hand, growing competition may force companies to take back and refund excess products from their customers. On the other hand, used product take–back and recovery is an important element for building up a 'green' profile, which companies are increasingly paying attention to. Today, most companies emphasize their reuse and recycling activities in environmental reports. While improving a company's environmental image alone may not be a sufficient justification for Reverse Logistics initiatives, it may account for additional benefits on top of other, economic advantages. Finally, taking back used products may be seen as a service element, taking care of the customer's waste disposal needs.

Third, environmental *regulation* is another reason for Reverse Logistics that is of growing importance. As sketched in the introduction, extended producer responsibility has become a key element of public environmental policy in several countries. In this approach manufacturers are obliged to take back and recover their products after use in order to reduce waste disposal volumes. While these legislative initiatives are mainly found in Europe, and partly in Eastern Asia, they surely have a world–wide impact in view of today's global markets.

Fourth, we mention *asset protection* as another motive for companies to take back their products after use. In this way, companies seek to prevent sensitive components from leaking to secondary markets or competitors. Moreover, potential competition between original 'virgin' products and recovered products is avoided in this way.

The different motivations driving 'reverse' goods flows have important implications for managing the corresponding logistics activities. In particular, they give indications as to which party initiates the flows. In the case of economically driven flows one can expect a more active role of the receiving party and hence a tendency towards a demand–pull situation. In contrast, legislation and commercial motivations may lead to a supply–push setting where the receiving party is mainly forced to respond to its customers' behaviour.

Dispositioning options

The above drivers are closely linked with the available options for recovering value from the goods under consideration; in other words, the dispositioning alternatives. In this context, many authors have adopted the categorisation proposed by Thierry et al. (1995) distinguishing different forms of recovery with respect to their re–entry point in the value adding process. Since this classification has often been discussed we do not need to elaborate on it in detail again and can therefore restrict ourselves to recalling the major terms.

First of all, items may possibly be *reused directly* without any major reprocessing except for cleaning or minor maintenance. Alternatively, *remanufacturing* conserves the product identity and seeks to bring the product back into an 'as new' condition by carrying out the necessary disassembly, overhaul, and replacement operations. In contrast, the goal of *repair* is to restore failed products to 'working order', though possibly with a loss of quality. *Recycling* denotes material recovery without conserving any product structures. Finally, returned products may not be reused at all. Therefore, we add *disposal*, in the form of landfilling or incineration, to the list of dispositioning alternatives.

It is clear that the available dispositioning alternatives have important logistics implications. They largely influence future 'forward' flows following the Reverse Logistics stage and therefore determine major boundary conditions. Aspects such as time–criticality, modes of transportation, and logistics integration with other flows may depend to a large extent on the dispositioning.

Actors
Furthermore, we would like to highlight the actors involved as another important dimension of Reverse Logistics flows. A major distinction can be made between products returning to a party in the *original supply chain* on the one hand, such as the original equipment manufacturer (OEM), and 'reverse' flows entering an *alternative chain* on the other hand. In the latter case a further distinction can be made between specialised parties relying exclusively on secondary resources and parties using them as additional input alternative. The former includes, e.g., specialised remanufacturing companies. As an example of the latter one may think of steel works making use of scrap metal. Again, there is a salient impact on the corresponding logistics processes. In particular, the configuration of Reverse Logistics actors sets important constraints for integrating 'forward' and Reverse Logistics activities.

Cycle time
Finally, we pay attention to the time period a product stays with its owner before entering a 'reverse' flow. As we see in the next section, cycle times for different types of Reverse Logistics flows differ considerably, varying from a few days in the case of reusable packaging to several years for end–of–life durable goods. The cycle time has a direct impact on feasible dispositioning options: in many cases the economic value of a good that is returned quickly may be expected to be higher than of a product that has stayed in the market for a long time. Furthermore, the cycle time largely influences logistics planning, namely appropriate forecasting approaches and opportunities for integrating 'forward' and 'reverse' flows.

3.2 Categories of Reverse Logistics Flows

Based on the above dimensions we can characterise a number of different categories of Reverse Logistics flows. Recall from Chapter 1 that we delineated Reverse Logistics as concerning the management of inbound flows of secondary goods opposite to the traditional supply chain direction. Within this scope we distinguish the following cases:

- End–of–use returns
- Commercial returns
- Warranty returns
- Production scrap and by–products
- Packaging

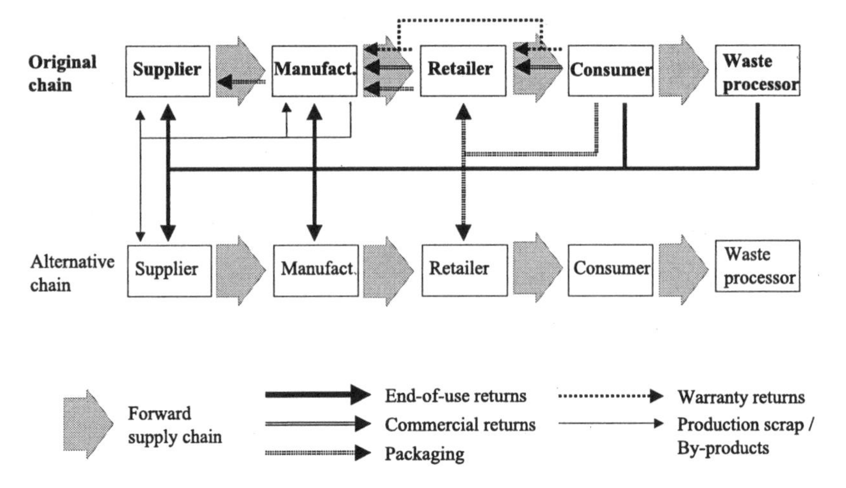

Fig. 3.1. Reverse Logistics flows in the supply chain

Figure 3.1 gives an overview of the different 'reverse' flows and their position in the supply chain. For each case, we indicate the former and the future product owner, i.e., the party that is responsible for the dispositioning decision. It should be noted that the arrows do not necessarily coincide with the corresponding physical flows. In particular, intermediates such as logistics service providers and subcontractors are not depicted here. The figure is meant as an illustration of the points where 'reverse' goods flows arise in the supply chain rather than as a description of reverse distribution channels. Furthermore, Table 3.1 characterises each of the above categories in terms of the dimensions addressed in the previous section and lists some typical examples. In the remainder of this section we discuss each category in more detail.

End–of–use returns
Probably the most prominent category of Reverse Logistics flows concerns end–of–use returns. It is mainly this group that has triggered the growing interest in Reverse Logistics in recent years. End–of–use returns denote flows of goods that are disposed of after their use has been completed. This group is in itself diverse and contains a wide range of examples. Alternatively, one may also find the term end–of–life returns for this type of flows. We prefer a somewhat broader perspective including returns of goods that have not necessarily reached the end of their technical or economic 'life'. In particular, we include returns of leased products in this group. End–of–use returns typically originate from consumers or waste processors. Moreover, the time between the original receipt and the return tends to be relatively long. Possible drivers for companies to deal with end–of–use return flows encompass the entire spectrum discussed in the previous section:—

First of all, used products may represent a valuable resource, which is economically attractive to recover. While direct reuse tends to be unfeasible in most cases, remanufacturing and recycling are the major recovery options for this category. Given the economic benefits, these kinds of end–of–use products may be attractive for both original manufacturers and specialised recoverers. The importance of market entry barriers, such as customer accessibility and required product knowledge for recovery, largely determine the relative advantages of both groups of actors. IBM's computer refurbishing, the copier remanufacturing initiatives discussed in the introduction, and the carpeting recycling projects are some of the numerous examples of the first case. Typical examples of the second case include tyre retreading by small independent shops and cellular telephone remanufacturing. As discussed above, economic motivations for end–of–use flows often go together with marketing drivers, namely the striving for an environmentally conscious company profile. We recall in this context the aforementioned example of Kodak's single–use cameras.

Another major group of end–of–use returns is due to environmental regulation. Typically, it is the original producer that is charged with the responsibility for environmentally sound end–of–life management in these cases. Although the actual processing of the returned goods may be outsourced to a third party the producer bears the financial responsibility. Given the lack of economic benefits and the legal restrictions to disposal, material recycling is the typical solution found in these cases. As typical examples one may think of the implementation of the Dutch electronics take–back legislation (see also Chapter 2) and scrap car recycling in Taiwan (Lee, 1997).

Finally, asset protection goals may also motivate end–of–use return flows. As discussed above, original equipment manufacturers may want to retrieve their products after use to prevent competitors from taking advantage of them. For example, this appears to be an important aspect in the collection of used toner cartridges by OEMs (see Chapter 1). In general, the collected used products may subsequently be recovered in the form of remanufacturing or recycling. However, they may also be disposed of after making sure that sensitive components and information have effectively been destroyed. The latter reflects current practice, e.g., for certain electronic components in the computer industry.

Commercial returns

Another important category of Reverse Logistics flows concerns commercial returns, denoting product returns undoing a preceding business transaction. In this case, the buyer returns products to the original sender against refunding. In principle, commercial returns may exist between any two parties in a supply chain that are in direct business contact. However, the most important cases concern returns from retailers to manufacturers and from consumers to retailers. Commercial returns primarily depend on the channel power of the different supply chain parties and represent a transfer of financial risk from

Table 3.1. Characteristics of different categories of Reverse Logistics flows

Description	*Cycle Time*	*Drivers*	*Dispositioning*	*Actors*	*Examples*
1. End–of–use returns					
products disposed of after completing use	long	economic, marketing	remanufacturing, recycling	original chain, alternative chain	electronic equipment remanuf., carpet recycling, tyre retreading
		regulation	recycling	original chain	White & Brown Goods Act (NL)
		asset recovery	remanufacturing recycling disposal	original chain	toner cartridges collection impairment of computer components
2. Commercial returns					
unused products returned for refunding	short, medium	marketing	reuse, remanufacturing, recycling, disposal	original chain	retailer overstocks of pc's, fashion clothes, cosmetics; catalogue retailers
3. Warranty returns					
defective or damaged products	medium	marketing, regulation	repair, disposal	original chain	defective household appliances, rotable spares
4. Production scrap & by–products					
production scrap and by–products	very short	economic, regulation	recycling, remanufacturing	original chain, alternative chain	pharmaceutical industry, steel works
5. Packaging					
packaging material and product carriers	short	economic	reuse	original chain, alternative chain	pallets, crates, bottles
		regulation	recycling	original chain	Green dot system (D)

the buyer to the seller. This is particularly relevant for products involving a high risk of obsolescence, for example due to seasonality (e.g. sun–care) or short product lifecycles (e.g. personal computers, fashion clothes). Commercial returns as such are not a new phenomenon. For example, catalogue retailers have always faced product returns from customers as an integral element of their business processes. However, returns are becoming increasingly important due to a concentration of market power, in particular in the form of large retail chains. A recent survey among US manufacturers reports in this context on return rates of up to 20% for personal computer manufacturers and of up to 30% for book publishers (Rogers and Tibben–Lembke, 1999).

Many dispositioning options for commercial returns are possible. Since the products are unused and, in general, not defective they may be reused, i.e. resold directly, possibly on an alternative market. However, this option may be highly time critical, in particular in the case of short product lifecycles (see also Chapter 2). Moreover, the occurrence of commercial returns in itself indicates a lack of market demand. Upgrading the returned products to new standards may be another alternative. Finally, material recycling or even disposal may be the last resort. In any case, commercial returns imply a financial disadvantage for the original seller.

Warranty returns

Warranty returns form a category of secondary goods flows contrary to the conventional supply chain direction that has been around for a long time. It refers to failed products that are returned to the original sender. This may concern products that failed during use but also goods that were damaged during delivery. Moreover, we also include rotable spare parts in this category. Finally, one may also think of product recalls due to security hazards. Warranty returns rely on both marketing considerations concerning customer service and on legislative rules. Repair is the typical dispositioning option for this category. However, disposal may be another alternative, possibly implying some refunding or replacement to the customer.

Production scrap and by–products

Yet another category of Reverse Logistics flows concerns production scrap and by–products. In many cases, excess material resulting, e.g., from cutting or blending is reintroduced in the production process. Similarly, off–specification products may be reworked to meet quality targets. As they save resources, these kinds of 'internal return flows' are economically driven. In other cases, reducing emissions may be required by environmental regulation, especially for hazardous materials. By–products are often transferred to alternative supply chains.

Packaging

Finally, packaging is another major class of Reverse Logistics flows. Crates and refillable bottles as well as pallets and reusable boxes are among the earliest and best known examples of Reverse Logistics. Reuse of these product carriers is economically attractive since they can often be reused directly without major reprocessing, except for cleaning. Moreover, they tend to be returned relatively quickly since they are only required for goods transportation and become available again after delivery. Reusable packaging may either be returned to the original sender, such as roll–containers from supermarkets to suppliers, or transferred to alternative parties. In many cases, reusable packaging is owned by logistics service providers who take care of the recollection. In view of its large contribution to waste disposal volumes, packaging material has also become a target of environmental legislation. One of the most prominent examples is the German 'Green Dot' system, obliging manufacturers to take back and recover their product packaging. Recycling, especially of plastic materials, is the major element of this system (Duales System, 2000).

Given their importance to recent developments in Reverse Logistics our focus in the sequel is mainly on end–of–use returns. However, the above discussion should make clear that there are more types of 'reverse' goods flows. Moreover, we have seen in the example of IBM in Chapter 2 that the management of different Reverse Logistics flows may be intertwined. Therefore, we keep the overall picture in mind and explicitly refer to it where appropriate.

3.3 Literature Review

This section provides a review of academic literature related to the field of the present investigation. As discussed in Chapter 1, the focus of our analysis within Reverse Logistics is on distribution and inventory management, as being logistics core elements. Literature on both of these areas is discussed in detail in the subsequent chapters in the course of the analysis. At this point, we take a broader perspective and consider literature complementing our research. For this purpose, we review scientific literature concerning general Reverse Logistics issues which mainly have a conceptual and structuring character. In addition, we discuss insights from business areas that have a direct interface with logistics and thereby determine its boundary conditions. To this end, we consider related literature on marketing channels, on the one hand, and on production and operations management, on the other hand. Given the direction of our research, the focus is on quantitative analyses. In addition, we include literature sources that provide points of reference by structuring major issues, and qualitative case descriptions that may serve as a basis for a quantitative analysis. Table 3.2 summarises the sources discussed below. It should be clear that we do not intend to provide an exhaustive list of references but rather aim at giving an up to date picture of major issues, results, and research activities relevant to our investigation.

3.3.1 General Reverse Logistics Issues

Monographs
At present, we are aware of four monographs that are dedicated to Reverse Logistics specifically. In addition, issues in Reverse Logistics have been addressed in detail in some recent PhD theses. The aforementioned White Paper of the CLM (Stock, 1992) is among the first publications systematically analysing the area of Reverse Logistics. Focus is on the role of logistics in waste reduction. The author identifies four major issues namely source reduction, recycling, substitution, and disposal and discusses their impact on each of the functional areas of procurement, transportation, warehousing, and packaging. Based on interviews with US companies and industry and governmental organisations he concludes that Reverse Logistics is yet a beginning development and that most companies adopted fairly reactive approaches driven by environmental regulation. The results of this study have been updated by Stock (1998). He places Reverse Logistics in the context of environmental management and emphasizes its potential corporate benefits due to cost reductions and customer value-added. Numerous examples of Reverse Logistics initiatives world–wide are given. Best practices are highlighted in case studies.

In a similar approach, Kopicki et al. (1993) address Reverse Logistics as an element of solid waste management. Logistics issues are discussed in the context of municipal waste recycling, reusable packaging, product take-back, and specialised third–party recycling services. In addition to a large number of illustrative examples, two detailed case studies are worked out that concern photo chemical recycling and a retailer's waste reduction initiatives, respectively. Collection and processing infrastructure, industrial partnerships, and performance measurements are identified as key aspects of Reverse Logistics programs.

Rogers and Tibben–Lembke (1999) take a slightly different perspective on Reverse Logistics by analysing companies' product take–back policies. Based on a survey among US companies competitive reasons are identified as a largely dominating factor. The authors then focus on cost reduction and value recovery opportunities for reducing the financial burden of product returns and emphasize the role of Reverse Logistics as a potential competitive advantage.

In addition, it is worth mentioning a number of recent PhD theses that have addressed issues in Reverse Logistics and product recovery management. Thierry (1997) has provided one of the first systematic analyses of operations management issues in a product recovery environment. Based on several business examples he lays out a general structuring of this field. Moreover, MRP–concepts and logistics network design are discussed specifically. The author emphasizes the high level of uncertainty as one of the major challenges intrinsic to product recovery management. We frequently refer to his analysis throughout this monograph (compare also Section 3.1).

Table 3.2. Summary of complementary literature

Area	**Conceptual/Structuring**	**Examples/Cases**	**Quantitative Models**
General Reverse Logistics Context			
monographs and PhD theses	Stock(1992,1998) Kopicky et al (1993) Rogers and Tibben–Lembke (1999) Thierry (1997)		van der Laan (1997) Krikke (1998)
literature reviews	Fleischmann et al (1997) Bras and McIntosh (1999) Moyer and Gupta (1997)		
overview of Reverse Logistics issues	Vandermerwe and Oliff (1991) Guide et al (1998) Carter and Ellram (1998) Tibben–Lembke (1999) Rosenau et al (1996) Flapper (1996) Kokkinaki et al (1999)	Thierry et al (1995) Clendenin (1997) de Koster and van de Vendel (1999) Ayres et al (1997) Gotzel et al (1999) IEEE (1999) Flapper and de Ron (1996,1999) Van Goor et al (1997)	
Marketing Channels			
	Guiltinan and Nwokoye (1975) Ginter and Starling (1978) Pohlen and Farris (1992) Fuller and Allen (1995)	Jahre (1995) Johnson (1998) Chandrashekar and Dougless (1996) de Koster et al (1999) Lee et al (1998)	Savaskan et al (1999) Savaskan and Van Wassenhove (1999) Emmons and Gilbert (1998)

continued below

continued from above

Production and Operations Management			
overview of issues	Lund (1984) Guide (2000) Nasr et al (1998) Gungor and Gupta (1999)	APICS (1998)	Flapper and Jensen (1998)
disassembly	Brennan et al (1994)	Spengler and Rentz (1996)	Johnson and Wang (1995,1998) Penev and de Ron (1996) Meacham et al (1999) Krikke et al (1998) Zeid et al (1999) Pnueli and Zussman (1997) Sodhi et al (1998)
MRP for product recovery		Krupp (1993) Panisset (1988)	Flapper (1994) Inderfurth and Jensen (1998) Inderfurth (1998) Clegg et al (1995) Gupta and Taleb (1994) Taleb and Gupta (1996)
scheduling remanufacturing operations		Guide and Ghiselli (1995)	Guide et al (1997a,b) Guide and Srivastava (1997,1999) Guide (1996)

Furthermore, Van der Laan (1997) has analysed several classes of mathematical inventory models in a remanufacturing context. His results are referred to in detail in Part III of this text. Finally, Krikke (1998) has addressed two main issues in product recovery management, namely dispositioning and logistics network design. His contributions are discussed in more detail in Section 3.3.3 below and in Part II.

Literature reviews

To date, a few literature reviews have been published that encompass research on Reverse Logistics. However, most of them have a different focus and include Reverse Logistics as one sub–aspect rather than a major theme on its own. Many of the sources referred to in this monograph are discussed in the review by Fleischmann et al. (1997) addressing Operational Research (OR) models in a Reverse Logistics context. The material is structured around the areas of distribution management, inventory control, and production planning.

Moyer and Gupta (1997) focus specifically on product recovery in the electronics industry as one of the most prominent areas in the recent developments in response to growing environmental concern. Their review encompasses a rich selection of literature on a conceptual description of product recovery, on environmental hazards of electronic waste, on product design issues, and on related marketing considerations. Moreover, general concepts are illustrated in the specific case of printed circuit board disassembly. Finally, a number of related OR models is discussed.

In addition, we would like to mention a review by Bras and McIntosh (1999). Taking a remanufacturing perspective, research concerning related design issues on a product, process, and organisational level is discussed. Emphasis is on the role of remanufacturing as a high–level form of value recovery, as opposed to material recycling. Literature is subdivided into descriptive and prescriptive work. Much attention is paid to the role of product design as a major determinant of remanufacturing possibilities.

Overview of Reverse Logistics issues

A paper by Vandermerwe and Oliff (1991) is among the earliest influential contributions providing a systematic analysis of the business implications of product recovery. The authors consider the shift from a linear 'buy–use–dump' perception of economy towards a concept of reconsumption cycles. They highlight emerging management tasks in the areas of research and development, manufacturing, and marketing. In particular, adjusting product design, setting up bi–directional logistics infrastructures, and developing appropriate sales channels are identified as key issues. More recently, Guide et al. (1998) have pursued this stream of research by presenting a detailed comparison between traditional and recoverable manufacturing environments. The paper discusses the impact on business functions including logistics, purchasing, and production planning. The authors emphasize the issue of uncertainty as one of the major challenges in product recovery.

Carter and Ellram (1998) investigate drivers and constraints determining a company's Reverse Logistics activities. Based on a literature study they identify regulation and customer preferences as major stimulating factors. At the same time, inferior quality of input resources and a lack of stakeholder commitment are found to be major obstacles to successful Reverse Logistics programs. Tibben–Lembke (1999) highlights the impact of Reverse Logistics on the costs relevant to a purchasing decision (total cost of ownership). He argues that Reverse Logistics may have both increasing and decreasing effects on various cost categories, including education, transportation, and maintenance. The author concludes that taking Reverse Logistics into account as a cost determinant is becoming increasingly important. Rosenau et al. (1996) discuss consequences of product recovery from an accounting perspective. The authors argue that reusable items, such as reusable packaging, should be considered assets rather than expensed items. A case study is presented, comparing ten US companies in the vehicle assembly industry. The authors conclude that investment oriented approaches, such as a net present value calculation result in significantly better decisions as to the use of reusable versus disposable packaging than the current practice of focusing on the pay–back period.

Flapper (1996) focusses on logistics aspects specifically. He considers collection and processing as major sub–areas within a reverse channel and distinguishes demand– and supply–driven activities. Supply uncertainty, multiple supply alternatives, and generation of by–products are stated as distinctive characteristics of Reverse Logistics as compared to traditional production–distribution environments. Kokkinaki et al. (1999) take a look at opportunities for electronic commerce solutions in Reverse Logistics. They emphasize the apparent match between the extended information requirement in a Reverse Logistics environment and the recent advances in electronic information technology. The authors discuss potential applications both for purchasing and sales. Considering current business examples they conclude that the combination of electronic commerce and Reverse Logistics is a promising yet largely unexplored issue.

Case studies

While many examples of recent Reverse Logistics initiatives have been pointed out, the number of related detailed case studies in academic literature turns out to be fairly limited. A notable exception concerns a well–known study on copier recovery (Thierry et al., 1995). This paper provides a systematic description of the steps in the implementation of a product recovery strategy, moving from a simple repair program to an extended hierarchy of recovery options. In a similar context, Clendenin (1997) reports on a business process reengineering approach to optimise Reverse Logistics channel activities at Xerox. De Koster and van de Vendel (1999) analyse return handling in the retail sector. They compare 9 companies in The Netherlands, divided into food and non–food retailers and mail–order companies. An overview is given of different return flow types and corresponding issues in transportation and internal handling. As a conclusion, the authors formulate guidelines when to integrate return handling with routine 'forward' activities and when to approach them separately.

A number of less detailed business examples from several industries, including electronics and tyres, has been discussed by Ayres et al. (1997). Furthermore, it is worth pointing at a recent survey–based study concerning Reverse Logistics activities of major companies in Germany (Gotzel et al., 1999). For a wider collection of examples we refer to the above monographs. Additional material can be found in proceedings of conferences such as the IEEE International Symposia on Electronics and the Environment (IEEE, 1999), and the International Seminar on Reuse (Flapper and de Ron 1996, 1999) and, in Dutch, in a regularly updated case collection (Van Goor et al., 1997).

3.3.2 Marketing Channels for Reverse Logistics Flows

The development of appropriate marketing channels and an efficient assignment of tasks to the different supply chain parties have been analysed since the early days of Reverse Logistics activities. As early as in 1975 Guiltinan and Nwokoye proposed a classification distinguishing four reverse channel types, namely (i) channels using traditional middlemen, (ii) channels involving secondary materials dealers, (iii) channels based on manufacturer–controlled recycling centres, and (iv) channels including joint–venture resource recovery centres. A comparison is given that focusses on collection and sorting, storage, and market communications as main channel functions. Collection volume is found to be a major critical success factor. Ginter and Starling (1978) confirm this conclusion and consequently see an important role for intermediates consolidating small volume flows from consumers into large volume supplies to recoverers. Pohlen and Farris (1992) have pursued this analysis. Rather than distinguishing a few channel types they state a set of reverse channel functions and potential actors, which may be assigned to each other in manifold ways. In addition to the above list of activities, transportation, compactification, and reprocessing are named as typical channel functions. The role of co–operation and a more systematic channel design, including location analysis, are stated as important research issues. Fuller and Allen (1995) have proposed a slightly different reverse channel typology, including loose, temporary networks. They highlight the role of public policy in providing appropriate conditions for the viability of reverse channels and conclude that long–term success of product recovery critically depends on partnerships among governments, businesses, and consumers.

Several authors have presented empirical studies illustrating and extending the above findings. Jahre (1995) analyses alternative channel structures for household waste collection and recycling. In particular, she pays attention to the tradeoff between sorting and combined handling of mixed material streams. She identifies population density and the variety of different materials as major impact factors. Johnson (1998) compares several networks for ferrous scrap recycling in the USA. He shows processing volume to be a main determinant of the power allocation among the supply chain members, namely metal working companies, scrap processors and steel mills. Chandrashekar and Dougless (1996) consider pricing mechanisms for selling surplus material. They point to the growing institutionalisation of recycling markets, as illustrated by the Chicago recyclables exchange, which enables long–term index–priced contracts. De Koster et al. (1999) describe the industry–wide recycling system set up in response to the recent electronics take back legislation in the Netherlands (compare also Chapter 2). The authors are critical about the financing mechanism relying on a fixed recycling fee charged to the consumer and about the lack of incentives for more advanced forms of product recovery. Lee et al. (1998) report on similar recycling systems in Taiwan for several product categories including packaging, tyres, batteries, and cars.

In addition, some quantitative approaches to analysing 'reverse' marketing channels have recently been proposed. Savaskan et al. (1999) develop a game theoretic approach to assessing the performance of alternative reverse channel structures. Specifically, they analyse an OEM product recovery system where collection is carried out by the retailer, the manufacturer or a third party, respectively. Considering the system's efficiency and the individual parties' profits, the authors find collection by the retailer to strictly dominate the other solutions. In addition, product recovery is shown to be a means for channel co–ordination. Savaskan and Van Wassenhove (1999) extend the above model to the case of multiple retailers. In this case, there is no strictly dominant solution. Collection by the manufacturer is shown to result in a higher recovery rate and in a higher cost efficiency of the system. On the other hand, collection by the retailers increases competition and provides an instrument of price differentiation to the manufacturer. Furthermore, several authors have addressed return policies in the context of supply chain contracts. In particular, commercial returns between retailers and manufacturers have been investigated. For example, Emmons and Gilbert (1998) present a game theoretic model for analysing the impact of commercial returns between retailers and manufacturers. It is show that return policies may be beneficial to both the manufacturer and the retailer. For a general discussion of supply chain contracts we refer to Tsay et al. (1998).

3.3.3 Production and Operations Management Issues

Overview of issues

The role of production and operations management in product recovery has received substantial attention during the past decade. In particular, much focus has been on remanufacturing, accounting for a significant industrial sector of its own right. In his seminal work Lund (1984) emphasizes the potential of remanufacturing for reconciling economic and environmental goals by exploiting added manufacturing value incorporated in used products. Guide (2000) characterises the main operations management issues and reviews state–of–the–art research. Based on a survey among US remanufacturers he identifies seven complicating factors, namely (i) return uncertainty, (ii) a potential imbalance between supply and demand, (iii) a need for disassembly, (iv) uncertain yields, (v) a need for Reverse Logistics, (vi) material matching restrictions, and (vii) uncertainty and variability in the processing steps. The author concludes that formal systems for planning and controlling remanufacturing operations are underdeveloped and largely absent in current practice. A similar conclusion is drawn in a study of the Rochester Institute of Technology (Nasr, 1998). Gungor and Gupta (1999) take a somewhat wider perspective, addressing product recovery in the context of environmentally conscious manufacturing. An extensive literature review is given, including issues in product design, collection, disassembly, inventory control, and scheduling. Finally, it is worth mentioning a review by Flapper and Jensen (1998), which focusses on rework as a specific form of product recovery and surveys literature on corresponding OR lotsizing and scheduling models.

A large variety of business examples illustrating the above issues can be found, e.g., in the proceedings of the APICS Remanufacturing Symposia (APICS, 1998). However, as in the previous sections, detailed case studies are largely lacking also for the production and operations management issues in product recovery.

In addition to these general approaches, a number of specific production and operations management issues in product recovery has been investigated in more detail. In particular, disassembly planning, modified MRP–concepts, and scheduling of remanufacturing operations appear to be areas that have seen active research efforts. We briefly address each of these areas.

Disassembly

A need for disassembly is one of the most salient aspects distinguishing many product recovery systems from a traditional manufacturing environment. Brennan et al. (1994) contrast assembly and disassembly operations and point out that they are not symmetrical to each other. In particular, disassembly planning has to cope with additional dependencies among multiple items.

Many authors have addressed the optimisation of the disassembly depth and sequence. Most approaches rely on a graph representation of the product structure. Computational challenges may arise due to large problem sizes. In this context, Johnson and Wang (1995,1998) have presented a network flow model for maximising the recovery profit for a given product, balancing component values and disassembly costs. Similarly, Penev and de Ron (1996) consider optimal 'cannibalisation' sequences releasing a number of preselected components from a given product. Meacham et al. (1999) extend these approaches to multi–product models involving fixed costs and common parts.

A column generation algorithm is designed to cope with large problem sizes. Krikke et al. (1998) develop a stochastic approach, taking into account uncertainty in the condition of a product and its components, which may affect the feasibility of recovery options.

Zeid et al. (1999) discuss an artificial intelligence approach to implementing disassembly optimisation. Keeping track of previous disassembly results is suggested as a means for overcoming problems due to a lack of accurate data. Pnueli and Zussman (1997) emphasize the link between recovery and product design. They show how information from a disassembly analysis can be used to eliminate weak spots in a product design in order to increase its end–of–life value. In a somewhat different context, Sodhi et al. (1999) consider material separation in bulk recycling processes. Rather than by discrete disassembly operations, shredded material fractions are separated in centrifuges or special baths, based on differences in specific weight. The authors develop a scheme for minimising the number of processing steps for separating a fixed number of materials from a given mix. Finally, an application of disassembly planning to the environmentally conscious dismantling of residential buildings appears to be worth mentioning (Spengler and Rentz, 1996).

MRP in a product recovery environment
The use of MRP concepts in a product recovery context is another issue that is receiving significant attention. Traditional MRP logic faces a number of difficulties in recovery planning. Specifically, the dependencies between components that are simultaneously released by disassembly and the choice between multiple supply sources (e.g. different returned products) cannot be handled adequately by a simple level–by–level top down approach as in traditional MRP. Therefore, several modifications to MRP have recently been proposed. Most of them rely on a 'reverse' bill of materials (BOM), documenting the recoverable subassemblies of a product and the processing times to release them. As not all components may be fully recoverable this 'reverse' BOM is not necessarily a symmetric picture of the original BOM.

Flapper (1994) addresses a situation where components for a final product may be obtained from the disassembly of used products as an alternative to purchasing new ones. Predetermined priority lists are used to deal with multiple procurement options for a required component. Inderfurth and Jensen (1998) and Inderfurth (1998) extend this model and analyse the issue of uncertain future availability of recoverable components in more detail. Specifically, reactive and proactive planning approaches are discussed. The issue of multiple supply alternatives is addressed more explicitly by Clegg et al. (1995). They propose a multi–period linear programming model for scheduling the disassembly of multiple used products and the reassembly of new and reusable components. Gupta and Taleb (1994) consider a situation with demand on a component rather than on a product level. They propose an MRP–algorithm for scheduling disassembly, taking into account dependencies between different components of the same product. Taleb and Gupta

(1997) extend this approach to a multi–product situation with parts commonality. To conclude, we refer to Krupp (1993) and Panisset (1988) for examples reporting on the practice of MRP for recovery planning in the automotive and railroad industries, respectively. Additional illustrative material can be found in the aforementioned conference proceedings.

Scheduling remanufacturing operations
Given the high level of uncertainty as one the main characteristics of remanufacturing, some authors are questioning the appropriacy of a purely deterministic concept such as MRP for this environment. A number of simulation studies has been presented that evaluates different scheduling policies for remanufacturing operations, including first–come–first–serve, due date oriented approaches, and batching (Guide et al., 1997a; Guide and Srivastava, 1997, 1999). The setting is motivated by the operations in an overhaul centre for military aircraft engines (Guide and Ghiselli, 1995). The authors conclude that the choice of the disassembly release strategy does not have a significant impact on system performance. Sophisticated batching or time–phased strategies fail, as a consequence of the varying processing requirements for each individual product. For the queue control at the work centres simple due date based priority rules are proposed. Guide et al. (1997b) have extended the above analysis by investigating the impact on capacity planning. The authors propose modifications to traditional rough cut capacity planning techniques by introducing discount factors, accounting for uncertain reusability and repair requirements. As an alternative to MRP, Guide (1996) proposes the scheduling of remanufacturing operations according to a drum–buffer–rope concept. Following the philosophy of synchronous manufacturing a continuous work flow is sought by focusing control on production bottlenecks.

This concludes our literature survey. We keep referring to the above issues throughout the subsequent analysis.

Part II

Reverse Logistics: Distribution Management Issues

4. Product Recovery Networks

4.1 Introduction to Reverse Distribution

Transportation of used or returned goods is probably the most salient issue in Reverse Logistics. Products need to be physically moved from the former user to a point of future exploitation or from the buyer back to the sender. In many cases transportation costs largely influence economic viability of product recovery. At the same time, it is the requirement of additional transportation that is often conflicting with the environmental benefits of product take–back and recovery. Therefore, careful design and control of adequate transportation systems is crucial in Reverse Logistics.

In a broader perspective, the above considerations point at distribution management issues in Reverse Logistics. In more traditional contexts distribution logistics has been structured in many ways, including internal versus external and inbound versus outbound transportation. In quantitative literature a distinction between distribution decisions on a strategic, tactical, and operational level is common. Corresponding decision models include location–allocation models, vehicle routing models, and dynamic routing and scheduling models (see, e.g., Crainic and Laporte, 1997). All of these issues can be expected to play a role also in a Reverse Logistics context: locations must be chosen for recovery facilities and collection points, product returns must be assigned to inspection or processing sites, and used products must be collected from former users or from specific take–back locations.

Considering examples from industrial practice, companies appear to experience novel Reverse Logistics issues especially on the strategic network design level. Recall in this context IBM's consideration of a European network for the recovery of used computer equipment (see Chapter 2). We discuss many more examples below. Given the apparent management concern, the focus of this part of the book is therefore on logistics network design. Other distribution management issues in Reverse Logistics are addressed at the end of this chapter.

More specifically, we concentrate on logistics networks for used product flows recovered on a product–, component–, or material–level. In this sense, we make a distinction with pure disposal networks and with returns of new products. In terms of Section 3.2 focus is on end–of–use returns. Other categories such as reusable packaging, by–products, and commercial returns are addressed to delineate the scope of our findings. As discussed in Chapter 1, Reverse Logistics is considered as a form of inbound logistics. Therefore, one may characterise the research object of this part of the book as the recoverer's logistic network design problem. Consequently, the scope of the analysis encompasses all supply chain stages for which this party assumes responsibility.

Traditionally, quantitative models have been developed to support the physical logistics network design, defining geographical locations, facilities, and transportation links. In Chapter 5 we propose an OR model facilitating analogous decisions in a product recovery context. To allow for an appropriate modelling, the goal of the present chapter is to characterise product recovery networks and to compare them with other logistics structures such as traditional production–distribution networks and waste disposal networks. Moreover, we look for a more detailed structuring of this field, distinguishing different types of product recovery networks. To this end, we take a broader perspective considering both geographic, economic, and organisational network aspects, namely the parties involved, their responsibilities, and the corresponding decision and control issues.

We base our analysis on a set of recently published case studies in literature and on direct contacts with industry, namely with IBM as discussed in Chapter 2. Each of the case studies includes a quantitative model and provides detailed information on the network considered. Bringing together these cases involving different industries appears in itself worthwhile since literature in this area is not yet well developed. Moreover, commonalities among the cases indicate general characteristics of product recovery networks. To understand the observed differences we introduce a set of potential factors influencing logistics network design. Positioning the available case studies in this setting, we identify a number of clusters of similar network characteristics and explanatory factors and in this way derive distinct product recovery network classes. We underpin our findings by considering additional examples and experiences from industry.

This chapter is structured as follows. In Section 4.2 we review case studies addressing logistics network design for product recovery. In Section 4.3 we bring the different examples together to identify common characteristics and compare them with other types of logistics networks. In Section 4.4 we derive a classification of product recovery networks. Introducing a general map of recovery context dimensions we reconsider the set of case studies to identify distinct network classes. In Section 4.5 we complement our analysis by taking a brief look at tactical and operational distribution management issues in Reverse Logistics, including vehicle routing aspects in particular.

4.2 Evidence from Current Practice

Recently, a considerable number of case studies has been reported in literature, all of which address the design of logistics networks in a product recovery context. In this section we provide a survey of these business cases. In each of the references a quantitative model for the network design problem is developed. Although we defer mathematical aspects to the next chapter, it appears that cases involving quantitative analysis provide a particularly valuable source of information since they describe the situation considered on a fairly detailed level. Many other, qualitative, papers exist but do not offer such comprehensive information. We use the latter to substantiate our findings.

For each case below we state the activities carried out in the network and the parties involved together with their responsibilities. Moreover, we mention the main Reverse Logistics drivers in each example. Finally, we pay attention to the network boundaries and links with external parties and other networks. The material presented in this section forms the basis for the analysis developed in the remainder of the chapter.

1. For the first case, we refer to the material in Chapter 2, illustrating network design issues arising in IBM's product recovery initiatives. As discussed, alternative locations have been considered for the inspection, dispositioning, and dismantling of used computer equipment.

2. Barros et al. (1998) report on a case study addressing the design of a logistics network for recycling sand resulting from the processing of construction waste in The Netherlands. While one million tons of sand used to be landfilled per year, reuse in large–scale infrastructure projects, e.g. road construction, is considered a potential alternative in line with environmental legislation. Therefore, a consortium of construction waste processing companies has investigated possibilities for establishing an efficient sand–recycling network. An important aspect to be dealt with is potential pollution of the sand, e.g. by oil. Therefore sand needs to be analysed before being reused. Three categories can be distinguished, namely clean sand that may be used without restrictions; half–clean sand, reuse of which is restricted to selected applications; polluted sand that needs to be cleaned after which it may be used freely. Cleaning of polluted sand requires installation of highly expensive treatment facilities. On the basis of these considerations a sand recycling network is to be set up, which encompasses four levels, namely crushing companies yielding sieved sand from construction waste, regional depots specifying the pollution level and storing cleaned and half–clean sand, treatment facilities cleaning and storing polluted sand, and infrastructure projects where sand can be reused. The locations of the sand sources, i.e. crushing companies, are known and their supply volume is estimated on the basis of historical data. As volume and location of demand are not known beforehand one has to resort to scenario–analysis. The optimal number, capacities, and locations of the depots and cleaning facilities are to be determined. The authors propose a multi–level capacitated facility location model for this problem formulated as a mixed integer linear pro-

gram (MILP) which is solved approximately via iterative rounding of LP–relaxations strengthened by valid inequalities.

3. Louwers et al. (1999) have considered the design of a recycling network for carpet waste. High disposal volumes (1.6 million tonnes of carpet waste landfilled in Europe in 1996) and increasingly restrictive environmental regulation on the one hand, and a potential of valuable material resources (e.g. nylon fibres) on the other hand has led the European carpet industry to considering a joint recycling network under the direction of DSM Chemicals. Through this network carpet waste is to be collected from former users and pre–processed to allow for material recovery. Since the content of carpet waste originating from various sources (e.g., households, office buildings, carpet retailers, aircraft and automotive industry) varies considerably, identification and sorting is required. Moreover, the sorted waste is to be shredded and pelletised for ease of transportation and handling. These pre–processing steps will be carried out in regional recovery centres from where the homogenised material mix is transported to chemical companies for further processing. The goal of the study is to determine appropriate locations and capacities for the regional recovery centres taking into account investment, processing and transportation costs. The authors propose a continuous location model. Using a linear approximation of the share of fixed costs per volume processed, all costs are considered volume dependent. The resulting nonlinear model is solved to optimality using standard software.
4. Carpet recycling has also been addressed in a case study in the USA by Ammons et al. (1997). The volume of 2.5 million tonnes of used carpet material landfilled per year makes recycling an economically interesting option. While the entire carpet recycling chain involves several parties, leadership is taken by DuPont, being a major producer of nylon fibres. A logistics network is investigated that includes collection of used carpeting from carpet dealerships, processing of collected carpet separating nylon fluff, other reusable materials and a remainder to be landfilled, and end–markets for recycled materials. Currently, the system is operational with a single processing plant. For alternative configurations the optimal number and location of both collection sites and processing plants are to be determined while delivery sites for recovered materials are assumed to be known. Moreover, the amount of carpet collected from each site is to be determined. Facility capacity limits are the main restrictions in view of the vast volume currently landfilled. The authors propose a multi–level capacitated facility location MILP to address this problem. They conclude that volume is a critical factor for the network layout.
5. Spengler et al. (1997) have examined recycling networks for industrial by–products in the German steel industry. Steel is produced from raw materials in several production facilities. The production of one ton of

steel gives rise to 0.5 tonnes of residuals that have to be recycled in order to comply with environmental regulation and to reduce disposal costs. For this purpose, different processing technologies are available. Recycling facilities can be installed at a set of potential locations and at different capacity levels, with corresponding fixed and variable processing cost. Thus, one needs to determine which recycling processes or process chains to install at which locations at which capacity level. Furthermore, one wants to optimise goods flows, under the assumption of linear transportation costs. The authors propose a modified multi–level warehouse location model with piecewise linear costs, which is used for optimising several scenarios.

6. Loosely related with a case study on copier remanufacturing Thierry (1997) has proposed a conceptual model for evaluating combined production/distribution and collection/recovery networks. The model addresses the situation of a manufacturing company collecting used products for recovery in addition to producing and distributing new products. Recovered products are assumed to be sold under the same conditions as new ones to satisfy a given market demand. The production/distribution network encompasses three levels, namely plants, warehouses, and markets. Products may be transported from plants to markets either directly or via a warehouse, yielding different transportation costs. Moreover, from each market a certain amount of used products needs to be collected. Subsequently, collected products are to be disassembled and tested on reusability, after which accepted products need to be repaired while rejected products are disposed of. These activities are carried out in the facilities of the 'forward' production/distribution network. For each facility a set of feasible operations and capacity restrictions are specified. Additionally, disposal sites are given. Disposal is feasible for all used products and is obligatory for products rejected after testing. In this model all facility locations are fixed externally. The model objective is to determine cost–optimal goods flows in the network under the given capacity constraints. Since facilities are given, no fixed costs are considered in the model. Decision relevant costs include variable production, handling, inspection, repair, disposal, and transportation costs. Since only variable costs are considered the problem is formulated as a linear program, which can be solved to optimality.

7. A similar situation has been addressed by Berger and Debaillie (1997). They propose a conceptual model for extending an existing production/distribution network with disassembly centres to allow for recovery of used products. Responsibility for product recovery lies with the original product manufacturer, who incurs all costs. The model is illustrated in a fictitious case of a computer manufacturer. The existing distribution network encompasses plants, distribution centres and customers. In the extended network used products need to be collected from the customers.

Collected products are to be inspected in a disassembly centre dividing them into three streams: high quality products can be repaired and shipped to a distribution centre for re–sale; products containing reusable parts may be disassembled and shipped to a plant to be reused in the production process; all other products are to be disposed of. Each plant and distribution centre can only use a limited amount of recovered products. While all facilities in the original network are fixed, the number, locations, and capacities of disassembly centres are to be determined. In a variant of this model the recovery network is extended with another level by separating inspection and disassembly/repair. After inspection, rejected products are disposed of while recoverable products are shipped to a repair/disassembly centre before entering a distribution centre or a plant. The authors propose multi–level capacitated MILPs to address these problems.

8. Jayaraman et al. (1999) have analysed the logistics network of an electronic equipment remanufacturing company in the USA. The company's activities encompass collection of used products (cores) from customers, remanufacturing of collected cores, and distribution of remanufactured products. Customers delivering cores and those demanding remanufactured products do not necessarily coincide. Moreover, core supply is limited. In this network the optimal number and locations of remanufacturing facilities and the number of cores collected are sought, considering investment, transportation, processing, and storage costs. The authors present a multi–product capacitated warehouse location MILP that is solved to optimality for different supply and demand scenarios.

9. Krikke et al. (1999) have reported on a case study concerning the implementation of a remanufacturing process at copier manufacturer Océ in the Netherlands. The recovery process can be subdivided into three main stages, namely (i) disassembly of return products to a fixed level, (ii) preparation, which encompasses the inspection and replacement of critical components, and (iii) re–assembly of the remaining carcass together with repaired and new components into a remanufactured machine. In addition, there are a number of supporting processes such as central stock keeping, sub–assembly production, and material recycling. While the supplying processes and disassembly are fixed, optimal locations and goods flows are sought for both the preparation and the re–assembly operations. Available options include two locations close to the main manufacturing site in The Netherlands and one more remote location in the Czech Republic allowing for lower personnel costs. Based on a MILP model the optimal solution minimising operational costs is compared with a number of pre–selected managerial solutions. Locating all processes in Prague appears to be optimal with respect to operational costs. However, this

solution also requires the highest investments and the differences in total relevant costs turn out to be fairly small.

10. Kroon and Vrijens (1995) have considered the design of a logistics system for reusable transportation packaging. More specifically, a closed–loop deposit based system is considered for collapsible plastic containers that can be rented as secondary packaging material. The system involves five groups of actors: a central agency owning a pool of reusable containers; a logistics service provider being responsible for storing, delivering, and collecting the empty containers; senders and recipients of full containers; carriers transporting full containers from sender to recipient. The study focuses on the role of the logistics service provider. In addition to determining the number of containers required to run the system and an appropriate fee per shipment, a major question is where to locate depots for empty containers. At these depots containers are stored and maintained, shipped to a sender upon request, and eventually collected from the recipient. Transportation of empty containers is carried out independently of the full shipment from sender to recipient, which may be realised by a different carrier. The expected volume and geographical distribution of demand is estimated on the basis of historical data concerning the number of shipments between given parties. Uncertainty is covered via scenario analysis. An additional requirement is balancing the number of containers at the depots. Since the total number of containers shipped from a depot during a planning period should equal the number of containers received, containers may be relocated among the depots. The decision problem is modelled as a MILP that is closely related with a classical uncapacitated warehouse location model.

4.3 Recovery Network Characteristics

We now analyse the above cases in order to derive a general characterisation of logistics networks for product recovery. Given our quantitative modelling goal we pay attention to physical network aspects especially. As a first step, common features of the presented examples are identified. Subsequently, a comparison is made with more traditional logistics networks.

4.3.1 Commonalities of the Surveyed Business Cases

In the terminology of Section 3.2 the above examples concern the management of end–of–use returns except for two cases dealing with reusable packaging (Kroon and Vrijens, 1995) and by–products (Spengler et al., 1997), respectively. All cases encompass a fairly similar selection of supply chain stages where the recoverer's responsibility begins with the collection of used products and ends with the distribution of recovered products. Consequently, the corresponding logistics networks span from a collection of actors releasing used products to another collection of actors with demand for recovered products. In the sequel these interfaces are denoted as disposer market, where the recoverer acts as a buyer, and reuse market, where the recoverer is a seller. While the specific steps in this transition differ per case the following groups of activities appear to be recurrent in product recovery networks:

- Collection
- Inspection / Separation
- Re–processing
- Disposal
- Re–distribution

We briefly describe each of these steps below. We remark that our structuring slightly differs from earlier approaches (e.g. Guiltinan and Nwokoye, 1975; Pohlen and Farris, 1992) by taking a logistics network perspective following the flow of goods. Therefore, we do not consider transportation and storage as distinct activities but rather as links between the above stages. In general, a transportation and a storage step may be required between each two of the above activities. Figure 4.1 gives a graphical representation of the activities within a product recovery chain together with traditional supply chain activities.

Collection refers to all activities rendering used products available and physically moving them to some point where further treatment is taken care of. Collection of used carpet from carpet dealerships (Ammons et al., 1997) and take–back of used copiers (Thierry, 1997) or computer equipment (see Chapter 2) from customers are typical examples from the above case studies. In general, collection may include purchasing, transportation, and storage

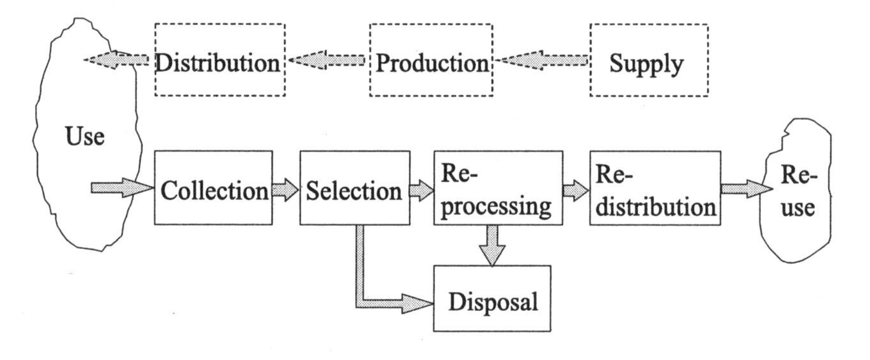

Fig. 4.1. The recovery chain

activities. Recall that collection may be motivated by different drivers, including economic benefits, marketing considerations, and legal obligations (see Section 3.1).

Inspection / Separation denotes all operations determining whether a given product is in fact reusable and in which way. Thus, inspection and separation results in splitting the flow of used products according to distinct reuse (and disposal) options. This applies, e.g., for distinguishing repairable and recyclable subassemblies of copiers (Krikke et al., 1999) and for inspection of sieved sand on pollution (Barros et al., 1998). Inspection and separation may encompass disassembly, shredding, testing, sorting, and storage steps.

Re–processing entails the actual transformation of a used product into a usable product again. This transformation may take different forms including recycling, repair, and remanufacturing (see Section 3.1). In addition, activities such as cleaning, replacement, and re–assembly may be involved. Examples are numerous, covering e.g. nylon recycling from used carpet (Ammons et al., 1997; Louwers et al., 1999), parts remanufacturing from used copiers (Thierry et al., 1995) or computers (Chapter 2) and cleaning of polluted sand (Barros et al., 1998).

Disposal is required for products that cannot be reused for technical or economic reasons. This applies, e.g., to products rejected at the separation level due to excessive repair requirements but also to products without satisfactory market potential, e.g., due to outdating. Disposal may include transportation, landfilling, and incineration steps.

Re–distribution refers to directing reusable products to a potential market and to physically moving them to future users. This may encompass sales (leasing, service contracts...), transportation, and storage activities. Sales of recycled materials (Ammons et al., 1997) and leasing of remanufactured copy machines (Thierry, 1997) are among the typical examples.

The similarities in activities are reflected in similarities in network topologies in the presented examples. Recovery networks can roughly be divided into three parts. See Figure 4.2 for a graphical representation. In the first part, corresponding to the collection phase, flows are converging from the disposer market typically involving a large number of sources of used products, to recovery facilities. Conversely, in the last part, corresponding to re–distribution, flows are diverging from recovery facilities to demand points in the reuse market. The structure of the intermediate part of the network is closely linked with the specific form of product recovery. In case of a limited set of processing steps carried out at a single facility, as in the examples of reusable packages (Kroon and Vrijens, 1995) and carpet waste pre–processing (Louwers et al., 1999), this network part may consist of a single level, comprising one or more parallel nodes. On the other hand, a complex sequence of processing steps involving several facilities may entail a multi–level structure of this network part including multiple interrelated flows. The latter case ap-

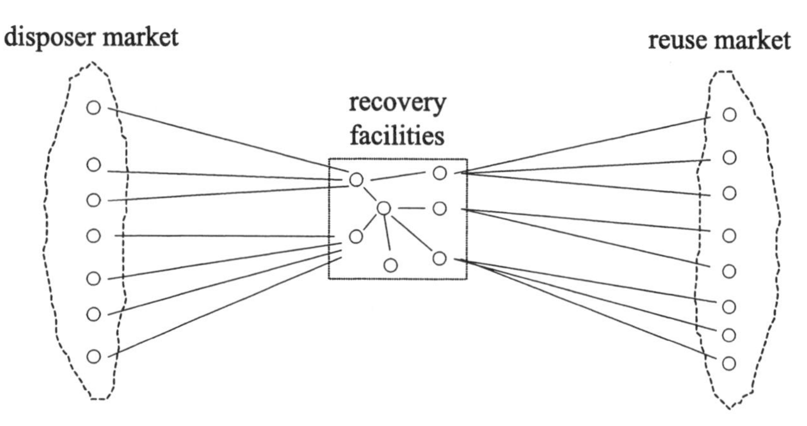

Fig. 4.2. Product recovery network topology

plies, e.g., to several remanufacturing examples (Krikke et al., 1999; Thierry, 1997). We discuss these differences in more detail in Section 4.4.

It should be noted that only the first part of a product recovery network actually concerns 'reverse' goods flows as defined in Chapter 1. In this part flows are directed from users to producers and undo steps of the original value chain. Subsequently, value is added and products move from a producer (recoverer) to a user just as in the traditional supply chain. To avoid misunderstanding, we therefore use the term 'product recovery network' rather than 'reverse logistics network'. In fact, the above cases emphasize again that Reverse Logistics should not be addressed in isolation but within the context of preceding and succeeding 'forward' flows.

In accordance with the general Reverse Logistics perspective as discussed in Chapter 1, it is the party carrying out the recovery process that is concerned with the logistics network design in all of the above examples. Determining the number and location of recovery facilities is a central task in the network design problems described above. In almost all cases geographical distribution and volume of both supply and demand are considered as exogenous variables. This gives product recovery networks a transshipment character. Sources and sinks are fixed while intermediary nodes are to be specified. We remark that sources and sinks, i.e. disposer market and reuse market, may coincide. Consider, e.g., reuse of containers (Kroon and Vrijens, 1995) and of office equipment (Thierry, 1997). In this 'closed loop' case, interaction between collection and re–distribution may add complexity to the network design problem. We discuss differences between 'closed loop' and 'open loop' networks further in Section 4.4. Furthermore we note that take–back obligations due to environmental legislation and 'green' market pressure often result in a supply 'push' situation. That is, availability of used products that need to be taken care of trigger the sequence of events rather than end product demand (Barros et al., 1998; Louwers et al., 1999; Thierry, 1997). At the same time, time restrictions tend to be weaker, in general, for collection than for distribution.

It has often been claimed that a high level of uncertainty is characteristic of product recovery management (see, e.g., Thierry, 1997). The above case studies support this vision with respect to network design issues. Demand for recovered products and materials appears to be difficult to forecast in many cases, the more so since reuse markets have only been emerging recently and often are not yet well established. Even more important, though, the availability of used products on the disposer market involves major unknown factors. In general, timing and quantity of used products coming free are determined by the former user rather than by the recoverer's requirements. Reliable planning of collection and recovery may therefore be a difficult task. Furthermore, the form of recovery and the sequence of processing steps required is often dependent on the quality of the input, e.g., pollution, damage, material mix, which is another unknown factor. We conclude by noting that uncertainty in the disposer market is particularly relevant in combination with a supply push, i.e., collection obligations.

4.3.2 Comparison with Other Logistics Networks

Having characterised product recovery networks we now compare them with logistics networks in other contexts. In particular, we consider traditional production–distribution networks. We start by noting that product recovery networks encompass several supply chain stages. In this sense product recovery fits well in the mindset of supply chain management, advocating co–ordination of the entire supply chain rather than considering single stages independently (see, e.g., Tayur et al., 1998). Roughly speaking, product recovery networks correspond to distribution networks encompassing supply, production, and distribution stages (compare Figure 4.1). The major differences between both contexts appear at the supply side. In traditional production–distribution systems, supply is typically an endogenous variable in the sense that timing, quantity, and quality of delivered input can be controlled according to the system's needs. In contrast, as pointed out in the previous subsection supply is largely exogenously determined in product recovery systems and may be difficult to forecast. Hence, supply uncertainty in a wide sense appears to be a major distinguishing factor between product recovery and traditional production–distribution networks.

As a direct consequence, traditional production–distribution networks typically do not include an 'inspection' stage similar to product recovery networks. Destinations of goods flows are, in general, known beforehand with more certainty as compared to the quality dependent processing routes in product recovery. While there may be exceptions, e.g., in the case of by–products or re–work, this is not the major focus of traditional production–distribution networks. Therefore, network structures may be more complex for product recovery, including more interdependencies. Another element that may render recovery networks more complex than traditional production–distribution networks is potential interaction between collection and (re–)distribution, e.g., combined transportation in closed–loop networks. We recall, however, that network complexity depends on the specific recovery process and may vary considerably per example. Finally, the number of sources of used products tends to be fairly large as compared to the number of supply points in a traditional setting. Bringing together a high number of low volume flows therefore appears to be characteristic of product recovery networks in particular.

On the distribution side differences between traditional and product recovery networks appear to be rather small. Possibly, demand uncertainty may be somewhat more prominent in the latter case since reuse markets are not yet well established and professionalisation tends to be lower. However, it can be expected that this distinction gradually disappears with product recovery becoming a 'normal' business. To a lesser extent this last observation may also hold for the issue of supply uncertainty. Co–operation agreements on the one hand and modern information technology such as tracking and tracing, machine sensoring, and electronic data interchange (EDI) on the other hand may contribute to a more stable environment for product recovery reducing, though surely not eliminating, supply uncertainty. As a general tendency, one may expect logistics networks for product recovery and for production–distribution to become more similar in the future, with product recovery becoming a standard supply chain element.

Similarly, it is worth considering the relation between product recovery networks and waste disposal networks. Disposal networks provide the logistics structure for collection, processing, and disposal of discarded products in the form of landfilling or incineration. We refer to Jahre (1995) for a detailed discussion. There are obvious analogies between disposal and recovery networks with respect to the 'supply' side. Used products need to be collected from many, possibly widespread sources and to be consolidated for further processing and transportation. Major differences between both network types arise on the 'demand' side. While a flow of recovered products is directed towards a reuse market, waste streams eventually end at landfill sites or incineration plants. The number of these disposal sinks is typically much smaller than the number of demand points in a reuse context. Hence, the divergent structure of the downstream network part is less prominent for disposal. Moreover, selection of disposal options is less sensitive to qualitative variations of the input. While waste streams may be sorted and split to some extent (e.g., material separation, removal of hazardous materials) according to different feasible disposal options (e.g., open or protected landfilling, incineration) these steps do not depend critically on the specific quality of discarded products. Hence, a considerably lower impact of input uncertainty is one of the major distinctions between disposal networks and networks for product recovery. However, it is worth noting that the line between both systems may not always be very sharp and that intermediate network types exist such as, e.g., for recycling of flue gas cleaning residues (Hammerschmid, 1990).

Finally, we take a look at commercial returns as a Reverse Logistics flow category concerning new rather than used products. Logistics network design for commercial returns and complaint handling has been the subject of a recent case study considering a large manufacturer of electronic household appliances (Anonymous, 1998). A major supply chain reengineering phase is carried out with the aim of reducing costs and stock levels and increasing flexibility. For this purpose, a national organisation of the supply chain is replaced by a structuring along larger geographical regions. Therefore, national distribution centres of several countries are being integrated into one central warehouse, supplying retailers in all of the countries concerned. Two categories of product return flows from retailers to the manufacturer are considered in this context. The first category concerns commercial returns, i.e. excess stock for which return rights are contractually granted. The second group concerns the reverse flow of products due to complaints related to the physical distribution process, e.g., incorrect delivery or damage. To date, all product returns in the above example are shipped to the national distribution centres where a classification is made of the product quality. Three options are available for further handling: A–quality products are added to the commercial stock at the national warehouse, B–quality products are sold in personnel stores, and the remainder of the products is scrapped or recycled externally. In addition to the product classification, investigation of complaint reasons and responsibilities is taken care of at the national distribution centre. The question arises how to integrate return flow handling in the new, supra–national supply chain structure. Two main options have been considered, namely concentrating all returns in the central European warehouse versus handling returns locally on a national basis. While the first approach complies better with the distribution network structure the latter may avoid unnecessary transportation of defective goods over long distances. Priority was eventually given to the first, centralised solution due to (i) coherence of complaint handling with the supply chain structure, (ii) avoiding local stocks that are difficult to control, and (iii) concentrating personnel and responsibilities. This example supports our vision outlined in Section 3.2 that commercial returns represent a financial burden in the first place. Therefore, a careful tradeoff is required concerning marketing benefits, such as an increasing market share, and return costs. In addition, organisational efforts to minimise return volumes by means of improved control and clear responsibilities appear to be more of an issue for these flows than logistics network optimisation. Therefore, it seems appropriate to specifically concentrate on used product recovery here rather than to address Reverse Logistics flows in general.

4.4 Classification of Recovery Networks

While we have identified a number of general characteristics of product recovery networks in the previous section, the networks encountered in the various case studies are surely not identical. Some discriminating factors such as network complexity and impact of uncertainty have already been mentioned in Section 4.3. In this section we consider distinctions within the class of product recovery networks in more detail. Main differences concerning the structure of the logistics networks in the above case studies refer to the following:

- degree of centralisation
- number of levels
- links with other networks
- open versus closed loop
- degree of branch co–operation.

Centralisation refers to the number of locations at which similar activities are carried out. In a centralised network each activity is installed at a few locations only, whereas in a decentralised network the same operation is carried out at several different locations in parallel. Centralisation may thus be seen as a measure for the horizontal integration or 'width' of a network. Analogously, the *number of levels*, referring to the number of facilities a goods flow visits sequentially, indicates the 'depth' or vertical integration of a network. In a single–level network all activities are integrated in one type of facility while in a multi–level network different activities are carried out at different locations. *Links with other networks* refer to the degree of integration of a new network with previously existing networks. A logistics network may be set up independently as an entirely new structure, or by extending an existing network. *Open versus closed loop* characterises the relation between incoming and outgoing flows of a network. In a closed loop network sources and sinks coincide so that flows 'cycle' in the network. An open loop network, on the other hand, has a 'one–way' structure in the sense that flows enter at one point and leave at another. Finally, the *degree of branch co–operation* relates to the parties responsible for setting up the network. Initiative may be taken by a single company, possibly involving subcontractors, or by a joint approach of an industry branch.

In Section 4.4.2 we characterise the networks considered in each of the case studies with respect to the above aspects. In order to explain the observed differences we take a broader perspective and analyse a set of context variables for each example. As a starting point, we introduce potential explanatory factors concerning the product recovery context in the next section.

4.4.1 Dimensions of the Network Context

To understand the observations concerning the network design in the individual case studies it is useful to consider each example in its broader context.

In what follows we structure the product recovery environment along three dimensions and briefly discuss attributes referring to products, markets, and resources, respectively.

- *Product* characteristics concern physical and economic properties of the used goods, such as weight, volume, fragility, toxicity, perishability, economic value, and rate of obsolescence. Obviously, each of these aspects influences the appropriate layout of a corresponding logistics network. In addition, we draw attention to the available recovery options, namely direct reuse, remanufacturing, and material recovery (see also Section 3.1). Again there is a clear link with the logistics network structure. In particular, the form of recovery determines the facilities required and hence the related investment costs. Moreover, as they result in different end–products different recovery processes correspond with different reuse markets. At the same time, feasibility of alternative recovery options depends itself on additional product characteristics, such as legal obligations and availability of status information. As an example for the latter think of product monitoring allowing for an optimal timing of replacement operations.
- *Market* characteristics refer to the different actors involved and their relationships. In general, suppliers, OEMs, service providers, independent recoverers, consumers, and public authorities all may play a role in setting up a recovery process. Interaction between the different parties has a major impact on the resulting supply chain structure and the corresponding logistics solutions. Each individual party chooses its responsibilities based on its relative power and on economic incentives. As discussed in the previous section, the recoverer sets up a logistics network spanning from a disposer market to a reuse market, in all of the above examples. However, the market conditions may differ significantly. On the one hand, the recoverer may be in the position to choose from the available supply on the disposer market. On the other hand he may be obliged to accept any product offered, e.g., due to legal obligations. Recall in this context the different Reverse Logistics drivers discussed in Section 3.1. Similarly, the recoverer may have a stronger or a weaker position on the reuse market which, in addition, may or may not coincide with the original product market. Furthermore, recovery may be managed by the OEM or an alternative company. In addition, individual tasks may be outsourced to third parties. As discussed in Section 3.1, this aspect largely influences opportunities for integrating Reverse Logistics activities with other logistics processes.
- Relevant *resources* influencing the recovery network design include recovery facilities, human resources, and transportation resources. In the above cases focus is on facilities, such as disassembly lines, test equipment, and recycling plants. Obviously, the required resources have a direct impact on the logistics network structure by largely determining the underlying economics. The relation between investment costs on the one hand and

operational costs on the other hand defines economies of scale, which are reflected in the degree of centralisation of the logistics network. Other resource related aspects of relevance include capacity restrictions and versatility. The latter is another important determinant of the aforementioned opportunities for integrating different logistics processes.

4.4.2 Product Recovery Network Types

We now bring together network properties and context variables in order to identify and characterise distinct product recovery network types. Table 4.1 lists for each case study a number of characteristics concerning both the logistics network and the recovery situation. The network properties follow those discussed at the beginning of this section. The recovery context is structured along the three dimensions products, markets, and resources as introduced in the previous section. The selection of aspects included in Table 4.1 is based on the information available from the case descriptions.

The cases can roughly be clustered in two groups having similar characteristics, namely Cases 2–5 on the one hand and Cases 1 and 6–9 on the other hand. Case 10 (Kroon and Vrijens, 1995) appears not to fit well in either group. Based on this observation and on general knowledge about other product recovery examples we propose to distinguish three types of product recovery networks, namely

- Bulk recycling networks (Case 2–5);
- Assembly product remanufacturing networks (Case 1;6–9);
- Reusable item networks (Case 10).

We note that this classification is process–oriented in the sense that it is the form of re–processing that is the major discriminating factor. A similar structuring has been proposed by Bloemhof–Ruwaard and Salomon (1997). Other studies have considered classifications based on the network initiators, e.g. manufacturer–integrated systems versus waste–hauler systems (Fuller and Allen, 1995; Ginter and Starling, 1978). These approaches appear to be more appropriate for a general analysis of product recovery systems, including organisational aspects, whereas our focus is on the logistics network structure more specifically. Although we do not claim completeness for the above list, we believe the proposed network types to cover many important cases. Of course, the classification is somewhat idealised and one may encounter mixed types in practice. We discuss each class in more detail below.

Bulk recycling networks
A first group of networks showing similar characteristics encompasses the examples of sand recycling (Barros et al., 1998), recycling of steel by–products (Spengler et al., 1997), and carpet recycling (Ammons et al., 1997; Louwers et al. 1999). All of these cases deal with material recovery from rather low value products. Disposer market and reuse market are different, in general, i.e.

the recovered materials are not necessarily reused in the production process of the original product. Consequently, material suppliers play an important role in these networks in addition to OEMs. Moreover, investment costs turn out to be very substantial in all of the above examples, due to advanced technological equipment required. In addition, the above cases share a rather centralised, open loop network structure involving a small number of levels. Finally, it is worth noting that the network is often established by relying on branch–wide co–operation.

Bringing the above aspects together we come to the following characterisation of bulk recycling networks. First of all, a low value per volume collected on the one hand and high investment costs on the other imply the need for high processing volumes. This conclusion is also supported by examples of paper recycling (Bloemhof–Ruwaard, 1996) and plastic recycling (Cairncross, 1992). Exploiting economies of scale is indispensable for making the recovery activities economically viable. Consequently, recycling networks tend to be highly vulnerable to uncertainty concerning the supply volume. The need for

Table 4.1. Product recovery network classes

Case	Recovery Network							Recovery Context												
								Products					Markets				Resources			
numbering from 4.2	centralised	multi–level	new network	extend existing network	open loop	closed loop	branch co–operation	low value	high value	complex structure	material reuse	parts / product reuse	OEM responsible	recovery mandatory	supply uncertainty	reuse in original market	high investment costs	high operational costs	high economies of scale	dedicated facilities
Recycling Networks																				
2	x	(x)	x		x		x	x			x			x			x		x	x
3	(x)		x		x		x	x			x		x	x	x		x		x	x
4	x			x	x			x			x						x		x	x
5	(x)	x	x		x		x	x			x		x	(x)			x		x	x
Remanufacturing Networks																				
1		x		x	x				x	x		x	x	(x)	x					(x)
6		x		x		x			x	x		x	x	x	x	x		x		
7			x		x				x	x		x			x					x
8		x		x		x			x	x		x	x	x	x	x		x		(x)
9	x	x		x		x			x	x		x	x	x	x	x		x		(x)
Reusable Item Network																				
10				x		x			(x)			x	(x)	(x)	x	x		x		

x = applies, (x) = applies partly

economies of scale is reflected by a centralised network structure. Moreover, co–operation within a branch may be an option to ensure high processing volumes. Scrap car recycling (Groenewegen and den Hond, 1993; Püchert et al., 1996) and household electronics recycling (Dillon, 1994) are additional examples of this approach. Co–operation is facilitated by an open loop character of material recycling, ensuring recovered material sales not to interfere with market shares in the original product market. Finally, a fairly simple network structure involving only a few levels results from the limited number of recovery options and the fact that technical feasibility of material recycling is not that critically dependent on the quality of the collected goods. Note, however, that input quality may be a major cost determinant, e.g., by influencing the purity of output materials.

Assembly product remanufacturing network
The examples of copier remanufacturing (Krikke et al., 1999; Thierry, 1997), cellular telephone remanufacturing (Jayaraman et al., 1999), printed circuit boards recovery (Berger and Debaillie, 1998), and used computer equipment recovery (see Chapter 2) form another group of networks having similar characteristics. All cases are concerned with reuse on a product or parts level of relatively high value assembly products. Recovery is often carried out by the OEM, and reuse and original use may coincide. Furthermore, supply uncertainty is reported to be an important factor in all of the above studies and operational costs for recovery appear to be relatively high. As for the recovery network, most of the above examples involve a fairly complex multi–level structure. Moreover, networks most often form a closed loop and rely on extending existing logistics systems.

From the above observations we draw the following conclusions concerning assembly product remanufacturing networks. Recovery of manufacturing added value is the main economic driver. Since the corresponding recovery activities, such as repair and remanufacturing, require (and reveal) intimate knowledge about the products concerned it is not surprising that they are carried out by the OEM in many cases. See Ferrer (1996, 1997) and Thierry (1997) for additional examples concerning the computer and automotive industries. However, if market entry barriers are low product recovery opportunities may also attract specialised third parties as, e.g., for tyre retreading (Ferrer, 1996) or recovery of toner cartridges (Scelsi, 1991). Product recovery has important marketing implications in these cases since markets for recovered products and original products may overlap. The latter also indicates a potential link between original logistics networks and recovery networks if the OEM is involved. Single–use cameras are an additional example (Ferrer, 1996). For these types of assembly product remanufacturing networks opportunities may arise for combining transportation or handling of both flows. A closed loop structure integrating both networks may therefore be a natural choice. Consequently, extending existing logistics structures may be a good starting point for the design of a recovery network.

Another important characteristic of added value recovery is a complex set of interrelated processing steps and options, which may entail a rather complex structure of the corresponding logistics network. This applies, in particular, to the intermediate network part between collection and re–distribution (see Section 4.3). Additional examples from the automotive and computer industries support this finding (see above). Moreover, feasibility of recovery options and the sequence of processing steps required depend strongly on the specific condition of the collected product, giving uncertainty a prominent role in remanufacturing networks. Decentralisation of certain activities such as testing and inspection may be one of the consequences for the logistics network layout.

Reusable item networks
Yet another type of networks can be found in systems of directly reusable items such as reusable packages. Although in literature we only found one comprehensive case study on logistics network design falling into this area (Kroon and Vrijens, 1995) there appears to be enough evidence to attempt a rough characterisation of this network class. As described in detail in Section 4.2 the above case considers a closed loop network for reusable packages. Upon return to a central provider responsible for the entire life cycle, packages can be directly reused. In this context timing of returns is reported to be an important element of uncertainty. Moreover, transportation and procurement of new packages are major cost factors. Finally, the logistics network has a decentralised, single–level structure extending a previously existing network.

We put these observations in a more general context as follows. Reusable items requiring only minor 'reprocessing' steps such as cleaning and inspection can be expected to lead to a rather flat network structure comprising a small number of levels, e.g., corresponding to depots. Moreover, a closed loop chain structure seems natural in this context since there is no distinction between 'original use' and 'reuse'. This applies, e.g., for many sorts of reusable packages such as bottles, crates, pallets (Bloemhof–Ruwaard and Salomon, 1997), plastic boxes (Trunk, 1993) and containers (Crainic et al., 1993). Determining the number of items required to run the system and prevention of loss are important issues in this closed loop situation (Goh and Varaprasad, 1986). Moreover, a fairly large number of reuse cycles and absence of other processing steps makes transportation a major cost component (Flapper, 1996). This may be a reason for a decentralised network including depots close to customer locations. Availability and service aspects point to the same direction. On the other hand, decentralisation renders balancing of item flows an important task in reusable item networks (see Crainic et al., 1993).

To conclude, we note that the line between 'reusable items' and more traditional items that are used multiple times is rather thin. The networks described above show much similarity with other closed loop systems such as, e.g., transportation fleet systems or video rental systems.

4.5 Vehicle Routing Issues

In addition to the strategic decisions considered in the previous sections, Reverse Logistics also gives rise to more tactical and operational distribution issues as explained in the introduction to this chapter. In this section we complement the above analysis of product recovery networks by briefly addressing related vehicle routing issues.

While determining vehicle routes and schedules is certainly an important task in Reverse Logistics it is not directly clear whether this is essentially different from other, more traditional logistics environments. Remarkably enough, literature on this issue is fairly limited. Figuratively speaking, moving things from B to A may not look much different from moving them from A to B. This is the more true if the transportation operation is outsourced. For a third party it may not matter much whether a specific shipment represents a 'forward' or 'reverse' movement in its customer's supply chain. In a recent study in the context of the Dutch electronics take–back legislation most carriers were found not to distinguish both cases in their planning (Romijn, 1999).

Yet some slight differences may be observed between collection and distribution, between inbound and outbound transportation. In particular, it has been pointed out that time pressure is often lower in the reverse channel. Picking up empty transportation packages such as reusable containers or pallets is less time–critical than full delivery shipments. Similarly, taking back disposed electronic equipment from a municipal collection site is less urgent than delivering new ones to a retailer, the more so since economic viability of many recycling systems relies on fixed disposal fees charged to new products rather than on the market value of the recoverable goods. In many cases, it is the accumulation capacity (and goodwill) of the sender which mainly restricts the timing of Reverse Logistics flows. For example, a supermarket can be expected to refuse used packaging to take up precious storage space. Similarly, collection sites, e.g., for disposed household appliances make use of containers, which need to be picked up once they are full. As a result of the weaker time constraints Reverse Logistics transportation can be expected to leave more room for efficient planning and optimisation than traditional 'forward' shipments. In this context, Jagdev (1999) reports on experiences concerning Reverse Logistics route planning at Burnham, a major US logistics service provider. He concludes that cost–revenue tradeoffs and vehicle loading time considerations are important aspects for determining efficient vehicle routes in a Reverse Logistics context, rather than relying on purely distance based planning. Another difference between collection tours and delivery tours concerns the number of stops. Given the large number of stops per tour, e.g., in public waste collection arc oriented planning approaches have been proposed in literature as opposed to traditional node oriented methods (compare Eiselt et al., 1995). However, applicability in a Reverse Logistics context seems limited. In most cases, as e.g. in the examples presented in

Section 4.2, the number of simultaneous sources of product returns is much smaller than for household waste collection and, in fact, even smaller than the number of demand locations for forward distribution.

All in all the above differences between forward and reverse channel vehicle routing appear to be rather limited. As for the network design issues discussed in the previous sections the major challenge may arise from the combination of both channels. Even if used products are returned to the original production location examples range from full integration of forward and return shipments on the one hand (e.g. reusable beer bottles) to complete separation on the other hand (e.g. toner cartridges returned via public mail services). Many reasons are possible for separating forward and reverse distribution even in a closed–loop situation, including different volumes, different handling requirements, and different timing. Moreover, integrating shipments may be complicated by vehicle loading restrictions. Rear–loaded vehicles often imply a first–in–first–out access to the load, such that collection and delivery stops cannot be mixed arbitrarily. On the other hand, the aforementioned timing flexibility in the reverse channel may facilitate integration of forward and reverse distribution. In the extreme case, planning of transportation routes may be completely forward flow driven, as in the example of reusable beer bottles, and return products are collected ad hoc along with the delivery tours.

In general, combined forward and reverse distribution gives rise to vehicle routing problems with delivery and collection stops. A similar situation has been addressed in a traditional logistics context for the combination of supply and delivery, e.g., in the grocery industry. So–called vehicle routing problems with backhauling have been formulated, for which several solution algorithms have been proposed (see, e.g., Toth and Vigo, 1999). Again the question arises whether this issue is substantially different for Reverse Logistics. Beullens et al. (1999a,b) argue that Reverse Logistics transportation problems are characterised by a relatively large fraction of pick–up customers as compared to more traditional settings and by a relatively large fraction of customers with both pick–up and delivery requests (denoted as exchange customers). The authors compare the impact of these parameters on the relative performance of different routing strategies, in particular separating versus combining delivery and pick–up tours. Based on a probabilistic analysis they conclude that the benefit of combined routing increases with the fraction of exchange customers and with the similarity of total delivery and collection volumes. Therefore, combined routing may prove particularly attractive in a Reverse Logistics context. The authors also emphasize the limitations to combined routing due to vehicle loading restrictions (see above). Expected benefits of combined routing may therefore be traded off against investments in special vehicle types allowing for more flexible access.

5. A Facility Location Model for Recovery Network Design

This chapter is concerned with quantitative decision models for an efficient design of logistics networks in a product recovery context. We start by reviewing literature in Section 5.1. Based on the characteristics identified in the previous chapter we then present a generic recovery network model in Section 5.2 and discuss its generality and limitations. Section 5.3 illustrates the model by means of two numerical examples. In particular, the impact of product return flows on the network design is highlighted. Section 5.4 presents a more systematic sensitivity analysis and investigates factors that determine network robustness. Finally, a number of model extensions is discussed in Section 5.5.

5.1 Recovery Network Design Models in Literature

Having characterised product recovery networks in the previous chapter we now address corresponding quantitative modelling issues: How can the properties of recovery networks appropriately be captured in quantitative models? How can the relevant tradeoffs in designing efficient recovery networks be captured mathematically to give quantitative decision support?

For logistics network design in a more traditional context, facility location models based on mixed integer linear programming (MILP) have become a standard approach. Numerous models have been presented in literature ranging from simple uncapacitated plant location models to complex capacitated multi–level multi–commodity models. At the same time, various solution algorithms have been proposed relying on combinatorial optimisation theory. We refer to Mirchandani and Francis (1989) and Daskin (1995) for a detailed overview of models and solution techniques.

Given this large body of research, MILP facility location models appear to be a natural starting point for recovery network design. The question then arises whether traditional models are flexible enough to include product recovery aspects. In other words, how do the differences between traditional production–distribution networks and recovery networks identified in Section 4.3 materialise in corresponding quantitative models?

As pointed out before, quantitative models have been presented in all of the case studies referred to in Chapter 4. Moreover, Realff et al. (1999)

have proposed an extended model related with the carpet recycling study at DuPont (Ammons et al., 1997). Finally, Marín and Pelegrín (1998) have developed an additional recovery network design model without a direct link to an industrial case. In this section we review these models, point out common characteristics and special features, and compare them with traditional facility location models from literature.

Table 5.1 summarises the major characteristics for each of the recovery network models. First, we indicate whether the integral network from disposer market to reuse market is modelled or only the 'reverse' network part from the disposer market to reprocessing. Second, we state whether the model explicitly takes into account a closed loop network structure or whether disposer and reuse market are modelled independently. Third, we consider the flow drivers for both the disposer and reuse market. A market–push from the disposer side indicates that whatever is released must be collected, whereas in a pull situation collection volumes may be decided upon, possibly up to a maximum available quantity. In the mathematical programming formulation these drivers are reflected in equality restrictions versus upper bounds, on the corresponding flow variables. Similarly, a market–pull from the reuse side refers to a given demand that must be satisfied, whereas sales volumes may be chosen in a push situation, again possibly up to some limit. Furthermore, we indicate the number of reuse options (dispositions) distinguished and how they are selected. Different dispositions may either have a fixed volume ratio or involve some more room for decision making. In addition, we list a number of model characteristics analogous with classifications of traditional facility location models, namely the number of network levels (for both fixed and free locations), the set of potential locations, the use of capacity restrictions (uncapacitated, capacitated, capacity to be selected), the number of time periods considered, and the number of inbound commodities distinguished. Finally, we state the type of mathematical programming formulation and the chosen solution method.

Table 5.1 summarises the major characteristics for each of the recovery network models. First, we indicate whether the integral network from disposer market to reuse market is modelled or only the 'reverse' network part from the disposer market to reprocessing. Second, we state whether the model explicitly takes into account a closed loop network structure or whether disposer and reuse market are modelled independently. Third, we consider the flow drivers for both the disposer and reuse market. A market–push from the disposer side indicates that whatever is released must be collected, whereas in a pull situation collection volumes may be decided upon, possibly up to a maximum available quantity. In the mathematical programming formulation these drivers are reflected in equality restrictions versus upper bounds, on the corresponding flow variables. Similarly, a market–pull from the reuse side refers to a given demand that must be satisfied, whereas sales volumes may be chosen in a push situation, again possibly up to some limit. Furthermore,

Table 5.1. Characterisation of recovery network models

	reverse / integral network	open / closed loop	disposer market driver	reuse market driver	# dispositions	dispositioning	# network levels (fixed + free)	potential locations	capacities	# periods	# inbound commodities	mathematical programming
Barros et al (1998)	int	open	push	pull	3	fixed	4 (2+2)	discr	cap	1	1	MILP
Louwers et al (1999)	rev	open	pull	push	N	fixed	3 (2+1)	cont	select	1	N	nonlin
Ammons et al (1997)	int	open	pull	push	N	fixed	3 (1+2)	discr	cap	1	1	MILP
Realff et al (1999)	int	open	pull	push	N	fixed	3 (0+3)	discr	select	N	N	MILP
Spengler al (1997)	int	open	push	pull	N	fixed	var	discr	select	1	N	MILP
Thierry (1997)	int	closed	push	pull	2	fixed	3 (3+0)	fixed	cap	1	1	LP
Jayaraman et al (1999)	int	open	pull	push	1	none	3 (2+1)	discr	cap	1	N	MILP
Berger and Debaillie (1998)	rev	closed	push	push	3	up. bd.	4 (3+1)	discr	cap	1	1	MILP
Krikke et al (1999)	rev	open	pull	pull	2	fixed	4 (2+2)	discr	uncap	1	1	MILP
Kroon and Vrijens (1995)	int	open	push	pull	1	none	3 (2+1)	discr	uncap	1	1	MILP
Marín and Pelegrín (1998)	int	closed	push	pull	1	none	3 (2+1)	discr	uncap	1	1	MILP

we indicate the number of reuse options (dispositions) distinguished and how they are selected. Different dispositions may either have a fixed volume ratio or involve some more room for decision making. In addition, we list a number of model characteristics analogous with classifications of traditional facility location models, namely the number of network levels (for both fixed and free locations), the set of potential locations, the use of capacity restrictions (uncapacitated, capacitated, capacity to be selected), the number of time periods considered, and the number of inbound commodities distinguished. Finally, we state the type of mathematical programming formulation.

Considering Table 5.1, most of the models can be characterised as single–period, multi–level, capacitated, discrete location models. All models are formulated as an MILP, with the exception of one continuous non–linear model (Louwers et al., 1999) and one pure LP model (Thierry, 1997). The number of network levels for which locations are to be determined varies between one and three. In addition, Spengler et al. (1997) consider general network flows between any two types of facilities, in order to model recycling process chains. Finally, Thierry (1997) proposes a pure allocation model, where all locations are fixed. Capacities for the various network activities, such as collection, storage, and processing may be fixed or be selected from a set of levels. Three models are uncapacitated (Kroon and Vrijens, 1995; Krikke et al., 1999; Marin and Pelegrin, 1998). Moreover, almost all models are stationary in the sense that they only consider a single time period. Only Realff et al. (1999) analyse network performance in a dynamic environment. While facility locations and capacities must be fixed for the entire planning horizon in this model, time–varying supply of recyclable material may lead to varying processing and transportation volumes. Furthermore, both single–commodity models addressing the recovery of one type of product and multi–commodity models distinguishing several types of input resources have been proposed. Finally, it is worth noting that almost all models are solved using standard commercial optimisation software. Only in two cases customised solution algorithms are developed (Barros et al., 1998; Marin and Pelegrin, 1998).

The above observations do not distinguish these models essentially from traditional facility location models. All aspects may also be encountered in models for a conventional production–distribution setting. In contrast, distinctive model properties arise as a consequence of novel market drivers. Conventional production–distribution networks are mainly demand–driven. Ideally, it is customer demand, which drives the entire supply chain. Consequently, traditional network design models typically consider demand to pull goods flows through the network. In other words, demand is the major exogenous factor to the decision problem. This is not true for recovery network design. As discussed in Chapter 4 recovery networks form a link between two markets and face exogenous factors also on the supply side. This is also reflected in the above models. Note that none of them is entirely pull driven but includes a supply–push either for the disposer market or the reuse market (or both). Mathematically this means that network flows may be subject to, possibly conflicting, constraints on the supply and on the demand side.

Another particular characteristic of the above models is due to the inspection and separation stage in recovery networks (compare Chapter 4). As explained before, product recovery in general involves multiple dispositioning options due to technical and quality restrictions. Therefore, most of the above models include a splitting of incoming return flows into different recoverable fractions. They may represent fractions of different materials in the case of recycling networks or different forms of recovery in remanufacturing networks (compare Chapter 4). If the different fractions arise in a fixed ratio, as in most of the above models, the model formulation is equivalent to multi–commodity network flows in a traditional sense. If fractions are only restricted by some bounds, as in Berger and Debaillie's model the situation is more complex. In this case, the optimisation problem involves an additional degree of freedom. The logistics network design and the form of recovery are then optimised simultaneously.

Finally, the interaction between forward and reverse channel in closed loop networks also gives rise to some model modifications. Thierry (1997) and Berger and Debaillie (1998) explicitly model facility sharing of both channels. In both examples the 'forward' distribution network is fixed. Thierry assumes that return flows are allocated to the existing 'forward' facilities for reprocessing and redistribution. Berger and Debaillie consider additional inspection centres to be set up for returns preprocessing. Subsequently, return flows are again allocated to existing facilities for further handling. In a more general perspective designing closed loop logistics networks may involve decisions as to which activities of the forward and reverse channel to integrate or separate. From a modelling perspective, one may think of this situation as a special type of multi–commodity flows that are oriented in opposite directions. Moreover, integration benefits due to economies of scales may require a more differentiated modelling of fixed cost terms. We return to this aspect in more detail in Section 5.5.

We conclude that current models for recovery network design are fairly similar to traditional facility location models, in particular to the class of multi–level warehouse location models. Major differences are due to additional flow constraints reflecting supply restrictions concerning the disposer market. The models include supply–push constraints rather than being entirely driven by a demand–pull. Other modelling modifications are due to multiple return flow dispositions and to possible interaction between forward and reverse channel. As a consequence, most of the models have a multi–commodity flow character.

In Chapter 4 supply uncertainty was identified as a main characteristic of recovery networks. Therefore, it is remarkable that uncertainty is not addressed explicitly in any of the above models. All models are purely deterministic and only treat uncertainty in a conventional way via scenario and parametric analysis. Given the prominent role of uncertainty in product recovery one may wonder whether this approach is appropriate or whether it primarily reflects the computational difficulty of stochastic models. Recently, Newton et al. (1999) have advocated a different road by proposing a model for robust network design. In subsequent sections we analyse the impact of uncertainty on recovery network design in more detail. To this end, we first formulate a generic model summarising the essential characteristics of the individual models discussed above.

5.2 A Generic Recovery Network Model

As a basis for a systematic quantitative analysis of recovery networks we now formulate a general network design model making use of the network characteristics identified in Chapter 4 and the models from literature discussed in the previous section. As explained above, currently available recovery network design models have much similarity with classical warehouse location models. Therefore, we start from the latter and proceed by incorporating specific recovery network characteristics.

For the general network topology we refer to Section 4.3.1. Recall that recovery networks form a logistics link between two markets, namely a disposer market where the recoverer collects used products and a reuse market where he sells recovered products (compare Figure 4.2). This defines the network boundaries. In addition, we consider three intermediate levels of facilities, following the groups of activities identified in Section 4.3.1. More specifically, we include disassembly centres where the inspection and separation activities are carried out, factories for the re–processing and possibly new production, and distribution warehouses. Moreover, we distinguish two dispositions for the collected goods, namely recovery and disposal. Recovery may not be feasible for all used products collected, which is revealed during inspection in the disassembly centres. The general structure of this network is displayed in Figure 5.1.

Within this framework, the network design problem considered in this chapter concerns deciding upon the number of facilities, their locations and the allocation of the corresponding goods flows. Analogous with traditional facility location models we can translate this problem into a MILP optimi-

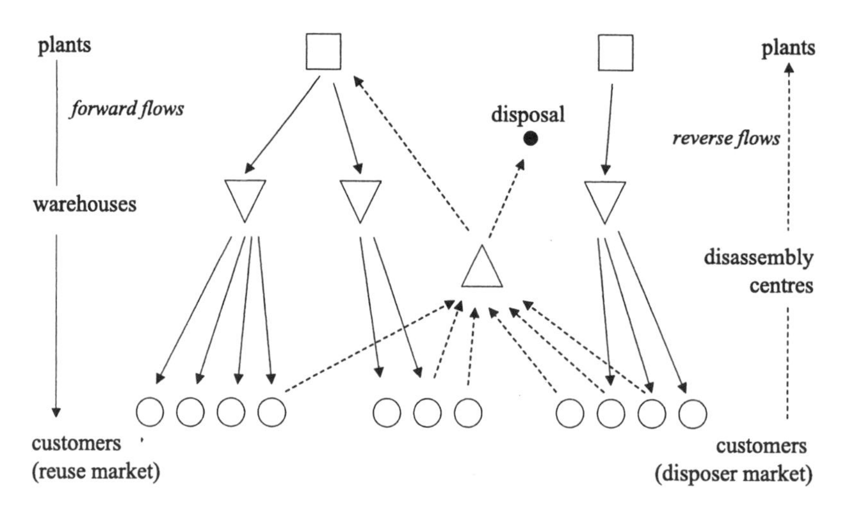

Fig. 5.1. Structure of the recovery network model

sation problem by modelling potential facility locations as binary and goods flows as continuous decision variables. The objective then is to minimise the sum of investment and operational costs. In achieving this goal, two main constraints need to be taken into account, in addition to logical conditions such as flow conservation and using opened facilities only. First, any solution must comply with the market conditions of both the disposer and the reuse market. Second, technical and economic restrictions of the dispositioning options must be met. We formalise the network design problem using the following notation.

Index sets

I = $\{1,...,N_p\}$ set of potential plant locations
I_0 = $I \cup \{0\}$, where 0 denotes the disposal option
J = $\{1,...,N_w\}$ set of potential warehouse locations
K = $\{1,...,N_c\}$ set of fixed customer loc. in disposer and reuse market
L = $\{1,...,N_r\}$ set of potential disassembly locations

Variables

X^f_{ijk} = forward flow: fraction of demand of customer k to be served from plant i and warehouse j; $i \in I, j \in J, k \in K$
X^r_{kli} = reverse flow: fraction of returns from customer k to be returned via disassembly centre l to plant i; $k \in K, l \in L, i \in I_0$
U_k = unsatisfied fraction of demand of customer k; $k \in K$
W_k = uncollected fraction of returns of customer k; $k \in K$
Y^p_i = indicator opening plant i; $i \in I$
Y^w_j = indicator opening warehouse j; $j \in J$
Y^r_l = indicator opening disassembly centre l; $l \in L$

Costs

c^f_{ijk} = unit variable cost of serving demand of customer k from plant i and warehouse j, including transportation, production, and handling cost; $i \in I, j \in J, k \in K$
c^r_{kli} = unit variable cost of returns from customer k via disassembly centre l to plant i; including transportation and handling cost minus production cost savings at plant i; $k \in K, l \in L, i \in I$
c^r_{kl0} = unit variable cost of disposing returns from customer k via disassembly centre l, including collection, transportation, handling, and disposal cost; $k \in K, l \in L$
c^u_k = unit penalty cost for not serving demand of customer k; $k \in K$
c^w_k = unit penalty cost for not collecting returns of customer k; $k \in K$
f^p_i = fixed cost for opening plant i; $i \in I$
f^w_j = fixed cost for opening warehouse j; $j \in J$
f^r_l = fixed cost for opening disassembly centre l; $l \in L$

Parameters

d_k = demand of customer k in the reuse market; $k \in K$
r_k = returns from customer k in the disposer market; $k \in K$
γ = minimum disposal fraction

We then formulate the general recovery network design model (RNM) as follows.

$$\begin{aligned} min\ ! \quad & \sum_{i \in I} f_i^p\, Y_i^p + \sum_{j \in J} f_j^w\, Y_j^w + \sum_{l \in L} f_l^r\, Y_l^r \\ & + \sum_{i \in I} \sum_{j \in J} \sum_{k \in K} c_{ijk}^f\, d_k\, X_{ijk}^f + \sum_{k \in K} \sum_{l \in L} \sum_{i \in I_0} c_{kli}^r\, r_k\, X_{kli}^r \\ & + \sum_{k \in K} c_k^u\, d_k\, U_k + \sum_{k \in K} c_k^w\, r_k\, W_k \end{aligned}$$

subject to

$$\sum_{i \in I} \sum_{j \in J} X_{ijk}^f = 1 - U_k \qquad \forall k \in K \qquad (5.1)$$

$$\sum_{l \in L} \Big(\sum_{i \in I} X_{kli}^r + \boldsymbol{X_{kl0}^r} \Big) = 1 - W_k \qquad \forall k \in K \qquad (5.2)$$

$$\boldsymbol{\gamma \sum_{i \in I_0} X_{kli}^r \leq X_{kl0}^r} \qquad \boldsymbol{\forall k \in K, l \in L} \qquad (5.3)$$

$$\boldsymbol{\sum_{k \in K} \sum_{l \in L} r_k X_{kli}^r \leq \sum_{j \in J} \sum_{k \in K} d_k X_{ijk}^f} \qquad \boldsymbol{\forall i \in I} \qquad (5.4)$$

$$\sum_{j \in J} X_{ijk}^f \leq Y_i^p \qquad \forall i \in I, k \in K \qquad (5.5)$$

$$\sum_{i \in I} X_{ijk}^f \leq Y_j^w \qquad \forall j \in J, k \in K \qquad (5.6)$$

$$\sum_{i \in I_0} X_{kli}^r \leq Y_l^r \qquad \forall k \in K, l \in L \qquad (5.7)$$

$$Y_i^p, Y_j^w, Y_l^r \in \{0, 1\} \qquad \forall i \in I, j \in J, l \in L \qquad (5.8)$$

$$0 \leq X_{ijk}^f, X_{kli}^r, U_k, W_k \leq 1 \qquad \forall i \in I, j \in J, k \in K \qquad (5.9)$$

This formulation, indeed, very much resembles a conventional multi–level warehouse location model. Equations (5.1) and (5.2) express the market conditions and ensure that all customer demand and returns are taken into account. Inequalities (5.5) through (5.7) are the usual facility opening conditions and (5.8) and (5.9) specify the domain of each variable. However, there are some product recovery specific elements (marked in boldface) that should be noted. First, the variables X_{kl0} together with constraints (5.3) reflect the additional degree of freedom concerning the dispositioning decision as discussed in the previous section. Note that there is no demand corresponding to the flow $\sum_l \sum_k r_k X_{kl0}$. The inequalities enforce a minimum disposal fraction for each inbound flow at the disassembly centres, to comply with technical and economic (in-)feasibility of product recovery. Second, constraints (5.4) display the required co–ordination between supply and demand. For each plant, incoming flows may not exceed outgoing flows. A possible gap represents the production of new products. While expressing a standard flow conservation condition, the particularity of constraints (5.4) is their dependence on two sets of exogenous parameters, namely d_k and r_k, that need to be balanced. We discuss the impact of this constraint on the mathematical structure of the model in more detail in Section 5.4. It should be noted that the absence of a disposal option on the plant level does not limit generality. Any surplus can already be disposed of at the disassembly centres rather than being shipped on to a plant.

We remark that the flow variables in above formulation correspond to paths from the disposer market to the plants and from the plants to the reuse market, respectively. As in conventional multi–level location models an equivalent arc oriented formulation is obtained by introducing flow variables $X^f_{ij} := \sum_k d_k X^f_{ijk}$ expressing the goods flow between location i and j (and analogously for all other transportation links). We prefer the above formulation here since it can be shown to have a tighter LP–bound (analogous with a conventional multi–level warehouse location model, compare Erlenkotter, 1978), which improves solver performance. Moreover, some specific scenario analysis is facilitated as we explain in Section 5.4. On the other hand, it should not be overlooked that the above path oriented formulation may result in a fairly large number of (continuous) variables in the order of $|I| \times |J| \times |K|$.

To see that the RNM is, indeed, a fairly general model that captures many different recovery situations we compare the above formulation with the aforementioned recovery network design models from literature. To this end, Table 5.2 summarises the characteristics of the RNM analogous with Table 5.1 above.

First of all, it should be noted that the RNM encompasses several alternative market situations. Since the parameters d_k and r_k can be selected independently any customer may belong to the disposer market, the reuse market or both. In this way, both closed loop and open loop networks can be modelled: If $d_k \times r_k > 0$ then customer k belongs to both the disposer and the reuse market, which allows for closed loop flows. In contrast, $d_k \times r_k = 0$ indicates a distinction between both markets and entails an open loop. Furthermore, both push and pull drivers describing the economics of the disposer and reuse market can be expressed. Large penalty costs c^w_k result in small values of W_k and hence by (5.2) in a collection obligation. In contrast, collection driven by a demand–pull, e.g. due to production cost savings, is captured by

Table 5.2. RNM model characteristics

reverse / integral network	open / closed loop	disposer market driver	reuse market driver	# dispositions	dispositioning	# network levels (fixed + free)	potential locations	capacities	# periods	# inbound commodities	mathematical programming
integr	both	both	both	2	up. bds.	5 (2+3)	discr	uncap	1	1	MILP

setting $c_k^w = 0$ for all k. Similarly, through the value of c_k^u both a push and a pull approach to the end market for recovered products can be modelled.

As discussed above, the RNM distinguishes two dispositions of returned products, namely recovery and disposal. This appears to be sufficient to capture the inspection issue in product recovery and the fact that returned goods can, in general, not all be reused in the same way. It is easy to extend the model to more recovery alternatives. Since constraints (5.3) are inequalities rather than prescribing a fixed disposal fraction, the RNM also includes the recovery policy optimisation element discussed in the previous section. In this way, the impact of the logistics network design on alternative recovery policies can be taken into account. Recall that transportation costs may have a significant impact on economic viability of product recovery options. In this context, it should be noted that the 'disassembly centres' may refer to any form of inspection and separation installations rather than being restricted to mechanical disassembly in a strict sense. What is essential is that feasibility of recovery options for the individual products is determined at this stage. Similarly, 'disposal' may include any form of recovery that is outsourced to a third party, e.g., material recycling. We only require this flow to leave the network at the disassembly centres.

For the rest, the model is kept as simple as possible, adapting an uncapacitated, single–period, single–commodity formulation. While all of these characteristics are easy to extend to fine–tune the model to specific applications it seems that this does not add to the general understanding of product recovery networks. Finally, as discussed before, the focus of our investigation is on the modelling of Reverse Logistics situations rather than on algorithmic issues. Therefore, commercial solvers are used for treating the RNM. Examples in Section 5.3 show that solution times for this approach are acceptable for reasonable problem instances .

In spite of the flexibility of the RNM it is worth noting some aspects that are not taken into account in the above formulation. First of all, the RNM is purely deterministic, just as the above models from literature, and does not explicitly capture the uncertainty that is typical of many product recovery settings. We address this issue in detail in the next two sections. Second, being a static model the RNM does not include the dynamic aspect of gradually developing and extending a recovery network. We return to this point in Section 5.4, too. Maybe the most important limitation of the RNM concerns a limited distinction between new and recovered products. As discussed above, the gap in inequality (5.4) represents the volume of new production, alternative to product recovery. Subsequently, both product categories are treated as perfect substitutes, only distinguished by possibly different prices, which can be incorporated in the variable cost parameters. Distinguishing demand for new and recovered products requires a multi–commodity extension of the RNM formulation. Finally, it should be noted that the above RNM formulation does not take into account synergies in integrating forward and reverse processes. We discuss these and other model extensions in more detail in Section 5.5.

5.3 Examples

We can now use the RNM to analyse the economics of product recovery networks quantitatively. In particular, we can investigate the impact of the 'reverse' goods flows on the network design, which is relevant for several reasons. On the one hand, recovery networks are not set up independently 'from scratch' in many cases but are intertwined with existing logistics structures, in particular if products are recovered by the OEM. The question then arises whether to integrate collection and recovery with the original 'forward' distribution network or rather to separate both channels. To this end, it is important to know how much product recovery is restricted by the constraints that are implied by existing logistics infrastructure. This question is the more important since many companies have gone through a major redesign phase of their logistics networks recently, notably in Europe. Global logistics structures have replaced national approaches. However, in many cases product recovery has not been taken into account yet. Therefore, one may wonder whether product recovery requires another fundamental change in logistics structures or whether it can efficiently be integrated in existing networks. On the other hand, supply uncertainty has been identified as a major characteristic of recovery networks in Chapter 4. To assess the consequences of supply *uncertainty* for the logistics network design it is helpful to first analyse the impact of (deterministic) supply *variations*. In other words, how robust are recovery networks with respect to return flow variations? An answer to this question at the same time concerns modelling appropriateness. Hence, we will see more clearly whether a deterministic model such as the RNM appears adequate for recovery network design or whether more advanced approaches are required, such as stochastic or robust optimisation techniques.

In this section we illustrate the impact of goods return flows on logistics networks by means of two examples concerning copier remanufacturing and paper recycling, respectively. The examples are inspired by real–life industrial cases and parameters are chosen in a realistic order of magnitude. However, we do not pretend to model any specific company's business situation.

5.3.1 Example 5.1: Copier Remanufacturing

Our first example follows in broad terms the direction of several case studies on copier remanufacturing (see, e.g., Thierry et al., 1995; Ayres et al., 1997). As discussed earlier, major manufacturers such as Xerox, Canon, and Océ are remanufacturing and reselling used copy machines collected from their customers. To be considered for remanufacturing a used machine must meet certain quality standards, which are checked during an initial inspection at a collection site. Remanufacturing is often carried out in the original manufacturing plants using the same equipment. Machines that cannot be reused as a whole may still provide a source for reusable spare parts. The remainder is typically sent to an external party for material recycling. In this example, we focus on the remanufacturing and recycling/disposal options. As mentioned before, our model may be extended to include additional recovery options such as spare parts dismantling. However, this extension does not change the essence of our analysis.

We consider the design of a logistics network for copier remanufacturing in a European context. To this end, we assume that a copier manufacturer serves retailers in 50 major European cities (capitals plus cities with more than 500,000 inhabitants). Customer demand at each retailer is assumed to be proportional to the number of inhabitants of the corresponding service region. As a first step, we consider a 'traditional' situation without product recovery. In this case, we need to determine a standard 'forward' production–distribution network, i.e. determine locations for plants and distribution warehouses and allocate the resulting goods flows. We restrict the possible plant locations to the 20 capitals, whereas warehouses may be located in any of the 50 cities considered. Moreover, for the sake of simplicity we assume that all relevant costs are location independent. Table 5.3 summarises the parameter settings for this example (ignoring the return flow parameters for the time being.)

For this example the RNM formulation reduces to a standard 2–level warehouse location problem involving 70 binary variables, 50050 continuous variables, and 3550 constraints. We solve this problem with a standard MILP–solver of CPLEX 6.0 based on LP–relaxation. Solution time on an IBM RS6000 computer is in the order of one minute. The solid lines in Figure 5.2 show the resulting optimal forward network, consisting of one central manufacturing plant in Frankfurt and five regional warehouses in Frankfurt, London, Barcelona, Milan, and Belgrade. For the sake of clarity, flows to and from the customers are omitted. Each customer is assigned to the closest warehouse. Total costs for this solution amount to k€ 44,314.

Table 5.3. Parameter settings in Examples 5.1 and 5.2

Description	Parameter	Value	
		Example 5.1 Copier Remanufacturing	**Example 5.2 Paper Recycling**
Fixed cost per factory	f^p	5,000,000	1,500,000
Fixed cost per warehouse	f^w	1,500,000	500,000
Fixed cost per disassembly centre	f^r	500,000	125,000
*Transportation costs per km per product**		*(product volume in pieces)*	*(product volume in tons)*
plant—warehouse	c^{pw}	0.0045	0.040
warehouse—customer	c^{wm}	0.0100	0.080
customer—disass.centre	c^{mr}	0.0050	0.060
disass.centre—plant	c^{rp}	0.0030	0.030
raw material source—plant	c^{sp}	n.a.	0.160
demand per 1,000 inhabitants	$d_k/\#\text{inh.}$	10	0.3
return rate	$\lambda = r_k/d_k$	0.6	0.7
minimum disposal fraction	γ	0.5	0.1
disposal cost per product	c^d	2.5	0.5
cost savings (recovery - production) per product	c^r	10.0	100.0
penalty cost unsatisfied demand	c_k^u	∞	∞
penalty cost uncollected returns	c_k^w	∞	0.0

*aggregated cost coefficients in model formulation:

$c^f_{ijk} = c^{sp}t_i + c^{pw}t_{ij} + c^{wm}t_{jk}$

$c^r_{kli} = c^{mr}t_{kl} + c^{rp}t_{li} - c^r$

$c^r_{kl0} = c^{mr}t_{kl} + c^d$

where t_{xy} denotes the distance between locations x and y and t_x denotes the distance between location x and the raw material source

Let us now assume that product recovery is introduced as an additional activity, which has to be integrated into the existing forward network. Suppose that the return volume of used products amounts to 60% of the sales for each retailer. Moreover, due to environmental regulation and service considerations all returned products have to be collected. After inspection 50% of the returned products turn out to be remanufacturable while the remainder has to be sent to an external material recycler. To design the return network, locations for the inspection/disassembly centres and allocations of the return goods flows need to be determined. Note that this includes a dispositioning decision for the remanufacturable machines, which may but do not have to be reused. We assume that inspection centres can be located in any of the 50 cities. Other parameters are again summarised in Table 5.3. The design of the return network for fixed forward locations results in a MILP problem with 50 binary and 5350 continuous variables and 5401 constraints, which we solve again with standard CPLEX routines. The dotted lines in Figure 5.2 show the optimal return network, comprising six regional inspection centres located in Frankfurt, London, Paris, Valencia, Milan, and Budapest. More-

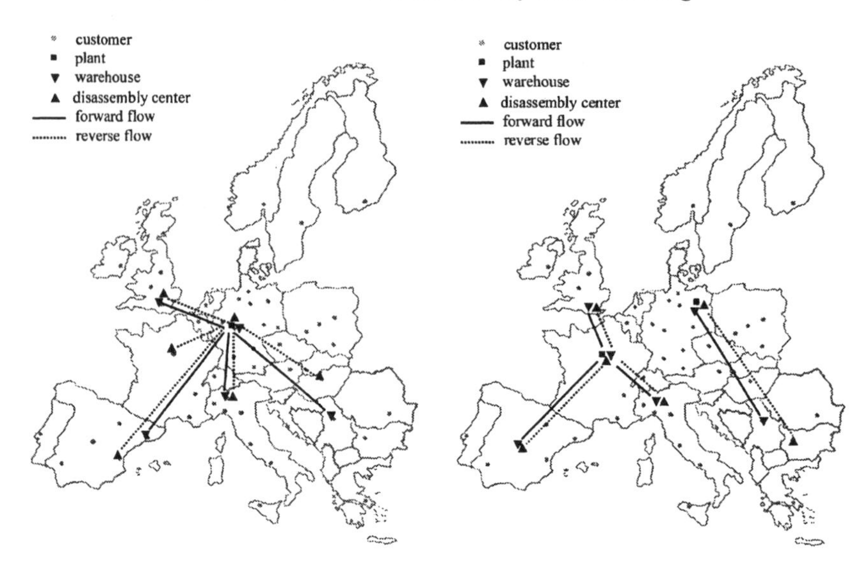

Figure 5.2: Optimal sequential network for copier remanufacturing

Figure 5.3: Optimal integrated network for copier remanufacturing

over, it turns out that all machines that are technically acceptable should actually be remanufactured. Total costs (including the forward network) are k€ 45,366. We see that forward and return network are very similar in this example. This may not be surprising since the degree of freedom for the return network design is fairly limited due to the fixed forward structure.

To assess the impact of this restriction let us now consider an integral design optimising both forward and return network simultaneously. We again use the parameters as in Table 5.3. The resulting MILP program now has 120 binary and 102,600 continuous variables and 8,620 constraints. Solving this problem in CPLEX requires about 10 minutes. Figure 5.3 shows the optimal integrated network for this example. It turns out that the optimal network now decomposes into two parts with manufacturing plants in Paris and Berlin, respectively. Clearly, the structure of this solution differs significantly from the network in Figure 5.2. Hence, we see that the product return flow can even change the optimal design of the forward network part. Due to the additional goods flows product recovery is a driver for decentralisation in this example. However, considering the cost effects puts this picture in a different perspective: total costs for the integrated solution amount to k€ 45,246, which comes down to savings of less than 1% with respect to the sequential approach. Hence, we conclude for this example that the sequential and the integrated recovery network design approach lead to different solutions but that cost differences are negligible. In other words, the fixed forward network structure does not impose significant restrictions on the design of an efficient return network. Clearly, this is good news for the manufacturer starting to engage into product recovery. Essentially the same results have been found in many other scenarios for varying input parameters. Before addressing this sensitivity analysis in more detail in Section 5.4 let us consider a second example.

5.3.2 Example 5.2: Paper Recycling

This case is motivated by the European paper recycling business. Waste paper comprises about 35% of total household waste volume in Europe. At the same time, increasing demand for pulpwood in paper production puts a heavy burden on forest ecosystems. Therefore, paper recycling has been a major issue for at least twenty years. As early as in 1975, Glassey and Gupta investigated maximum feasible recycling rates given the state of pulp and paper technology. They propose a simple LP model to determine production, use, and recovery of paper. Gabel et al. (1996) point out that the level of recycling also has important consequences for national economies by influencing geographical allocation of industrial activities. In this context, Bloemhof et al. (1996) studied the impact of mandated recycling quotas on the European paper industry. They show that forcing high levels of recycled content, taken as a measure to reduce Western Europe's solid waste problem, would severely hit Scandinavian industry. In view of the low population in the Nordic countries, these major pulp producers would have to import waste paper in order to produce recycled paper. Based on a LP network flow model the authors conclude that it is preferable both from an ecological and economic perspective to produce high quality paper, mainly containing virgin pulp, in Scandinavia while locating paper production with a high content of recycled pulp close to the population centres in Western Europe. Current observations from industry appear to confirm these findings (see Brown–Humes, 1999).

In this context, we consider the design of a logistics network for a European paper producer. Customers and potential facility locations are the same as in Example 5.1. However, we now have to take into account an additional cost element, namely raw material transportation. We assume that pulpwood is exclusively supplied from forests in Scandinavia and add its transportation as a location dependent element to the production costs. Moreover, we assume that transporting pulpwood is significantly more expensive than transporting paper. The last column of Table 5.3 summarises the parameter settings for this example.

Again, we first consider a pure 'forward' network without collection and recycling. Problem size and solution times are similar to Example 5.1. The bold lines in Figure 5.4 show the resulting optimal solution consisting of a central production plant in Stockholm and five regional warehouses in Stockholm, Hamburg, Zaragoza, Milan and Krakow. Total costs for this solution amount to k€ 19,570.

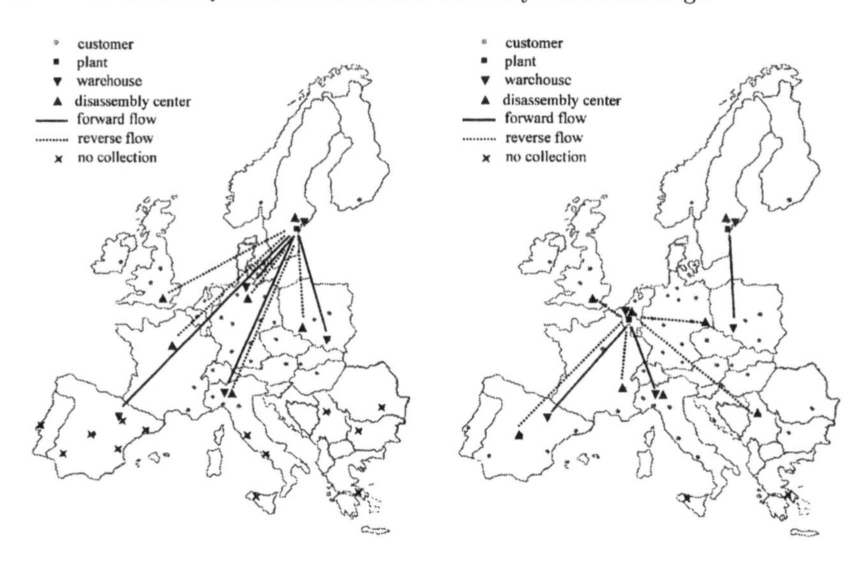

Figure 5.4: Optimal sequential network for paper recycling

Figure 5.5: Optimal integrated network for paper recycling

We now include recycling of waste paper. For this purpose, pre–processing centres need to be installed where collected paper is sorted and compacted and then transported to a production plant (compare Wang et al., 1995). In our model, processing centres play the same role as disassembly centres in Example 5.1. We assume that a maximum of 70% of the sales volume is available for collection at each customer. (For comparison note, e.g., that EU directives set minimum targets of recycled paper content for packaging material of 60%.) In line with current policy we assume that there are no take–back obligations for used paper. Hence, collection follows a pull approach. Residual waste paper volumes are assumed to be taken care of by competitors and local authorities. Finally, we assume that 10% of the collection volume is extracted at the pre–processing centres as being non–recyclable. The dotted lines in Figure 5.4 indicate the optimal collection network in this case. Six regional pre–processing centres are located in Stockholm, London, Paris, Milan, Hannover and Wroclaw. Moreover, due to the large distance from the processing site collection in southern Europe turns out not to be economically attractive, including the Iberian peninsula, southern Italy, and the Balkan. Total costs of this network (including the fixed forward locations) amount to k€ 17,990.

Finally, for this example also we consider an integral design optimising forward and return network simultaneously. Parameters are again as in Table 5.3. Figure 5.5 shows the resulting optimal solution. As in Example 5.1 the optimal network now decomposes into two parts. A plant in Stockholm now only serves the northern and north-eastern part of Europe, while all other countries are served from a new plant in Brussels. Note that this result is in accordance with what we observe in industry, as discussed at the beginning of this subsection. The collection strategy has also changed when compared to the sequential approach. With the exception of Athens and Palermo collection is now beneficial at all locations. As a consequence, the number of pre–processing centres has increased to eight. However, what is even more significant is that the total network cost decreased to k€ 14,540, which is about 20% lower than for the sequential design. Hence, in contrast with Example 5.1, optimising the forward and return network simultaneously not only leads to a different solution than a sequential approach but also results in a significant cost reduction in this case. The reasons for these different results are explained in the next section.

5.4 Parametric Analysis and Network Robustness

In order to understand the differences between the two examples presented in the previous section, we now analyse the impact of the return flows in the RNM more systematically. For this purpose, we first place this issue in a formal, mathematical context and reconsider our model from this perspective. Then we derive an explanation for our observation by analysing structural differences between the two exemplary cases and verify our hypotheses in additional numerical experiments. Finally, we apply our findings to the initial set of case studies from literature.

From a mathematical perspective, investigating the impact of the return flows on the network design comes down to a parametric analysis of the RNM with respect to the parameters r_k. Therefore, we can make use of the well developed theory of parametric mixed integer linear programming (see, e.g., Geoffrion and Nauss, 1977). Considering the MILP formulation introduced in Section 5.2 we see that each r_k occurs both in the objective function and in constraint set (5.4). This is a significant difference with traditional 'forward' uncapacitated facility location models which can be formulated such that (demand) volume parameters occur only in the objective function (see, e.g. Mirchandani and Francis, 1989). In the latter case the objective function is known to be piecewise linear and concave in the volume parameters and is therefore easy to compute on an arbitrary interval (Jenkins, 1982). For the recovery network it is the co–ordination of exogenous supply and demand represented by constraint set (5.4) that makes things more difficult. It is worth noting that these constraints, which couple the forward and the return network, somewhat resemble a capacity restriction for the recovery activities. In this sense, product return flows introduce a capacity issue into an otherwise uncapacitated network model. Using the arc oriented formulation sketched in Section 5.2 the RNM can be reformulated such that the parameters d_k and r_k occur only in the right–hand side. Therefore, the minimum cost function can be concluded to be piecewise linear in r_k for each k. However, it is not

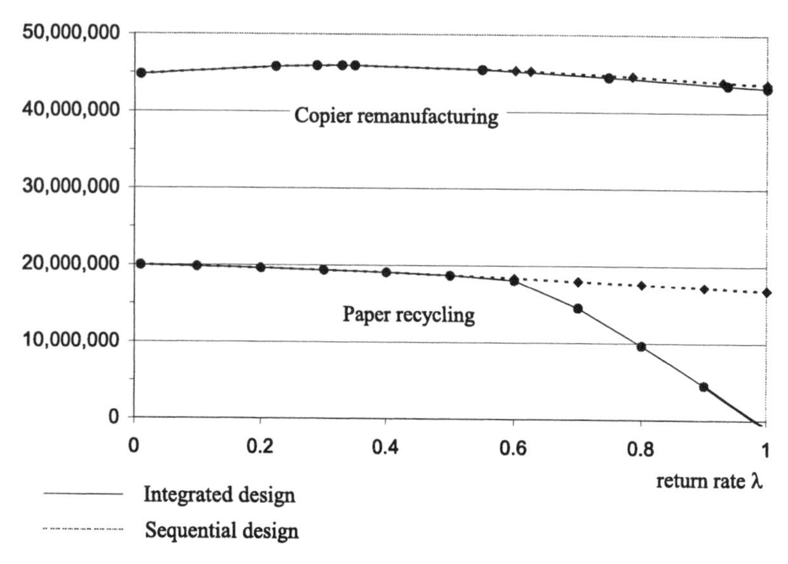

Fig. 5.6. Network costs as function of the return rate

necessarily concave. For computation we can use Jenkins' heuristic in this situation (Jenkins, 1982).

Figure 5.6 shows the minimum costs for Examples 5.1 and 5.2 as a function of the return rate $\lambda \in [0; 1]$ where $r_k = \lambda d_k$ for all k. The solid lines refer to the cost function of the integrated design optimising both forward and return network simultaneously, whereas the dotted line indicates the costs of the sequential approach. Not surprisingly, both approaches coincide for small return rates. For larger values of λ costs for both approaches differ, indicating that for these cases the return flows change the optimal design of the forward network. However, in the copier example the cost difference is negligible on the entire interval whereas costs for both approaches deviate significantly in the paper recycling example. (Note that the objective value may become negative since production and handling costs are not included and hence the cost advantage of product recovery is reflected in revenues.)

To explain the different impact of the return flows we consider the cost structures in both examples. First of all, it should be noted that the forward flows will, in general, dominate the optimal network structure since they are more important than return flows in terms of volumes, values, and time-criticality. Therefore, return flows can only be expected to influence the overall network structure significantly in the case of a major difference between the cost structures of the forward and return channel. In the electronics example geographical cost drivers are very similar for both channels. Demand and return volumes are distributed along the same geographical patterns and forward and return flows correspond with each other. Therefore, it is not surprising that optimal solutions for the forward and return network are also fairly similar and the impact of the returns on the overall structure is small. In contrast, there is an important difference between the cost elements of the forward and the reverse channel in the paper example. The structure of the forward network is dominated by costly raw material transportation from a fixed source on the boundary of the geographical area considered (i.e. forests in Scandinavia). In contrast, costs of the return network are independent of this source and are determined by the locations of the major customers (i.e. the population centres in Western and Central Europe). It is due to this difference in 'centres of gravity' that product recovery has a significant impact on the overall network structure in the paper recycling example. By substituting virgin input resources, recycling literally 'pulls' the network away from the original source towards the vicinity of the customers.

We have carried out a series of numerical experiments to test our argumentation and conclude that the similarity between supply and demand side both in terms of geographical distribution and cost structure is indeed a major determinant of the impact of product recovery on the overall network structure. We have varied parameters in the copier example over a large range without finding any case with a significant cost difference between the integral and the sequential design approach. This includes relaxing proportionality of returns and demand per customer, i.e., a non–uniform return rate. We have considered different return rates in different parts of Europe motivated, e.g., by regulation or customer attitudes (e.g. high return rates in Northern and Western Europe, intermediate return rates in Southern Europe and low return rates in Eastern Europe.) Still the cost deviation we observed between both design approaches was marginal.

We only found a relevant impact of product recovery on the overall network structure when including a major structural difference between forward and reverse channel as in the paper recycling example. However, even in this case, product returns do not always change the optimal forward network design. The economic incentive for product recovery is another important factor in this context. Lower production cost savings, lower penalty costs for not collecting returned products, and lower disposal costs all result in a smaller impact of the return flow since 'mismatching' returns can then be avoided altogether at low cost (to the producer). Finally, the number and uniformity of potential facility locations also appears to influence the cost deviation between the optimal integrated and sequential network design. Fewer potential locations tend to increase sensitivity.

We conclude that existing forward distribution networks do not form a barrier for setting up an efficient logistics structure for product recovery in many cases. Hence, product recovery can often be implemented efficiently without requiring major changes in existing production–distribution networks. Moreover, from a modelling perspective this means that forward and return networks may be addressed separately, which significantly reduces the

Table 5.4. Determinants of network sensitivity to return flow variations

Factor	Forward Network	Return Network
Geographical difference between disposer and reuse market	+	
Different cost structures of forward and reverse channel	++	
Incentives for product recovery	+	(+)
Few potential locations	(+)	+
High investment costs		+
Low minimum disposal fraction	(+)	+

++ = large impact on sensitivity
\+ = impact on sensitivity
(+) = limited impact on sensitivity

problem sizes. Care must be taken if forward and reverse channel differ largely with respect to geographical distribution and cost structure and return volumes are substantial.

Even if return flows do not have a significant impact on the forward network, the return part of the network may still be sensitive to changes in return volumes. In terms of the Figure 5.6 this refers to the changes in the slope of the minimum cost function. Sensitivity of the return network is an important aspect, e.g., when extending product recovery from a low volume activity to a larger scale. It should be noted that the situation is similar to traditional warehouse location models, for which a fairly robust behaviour with respect to moderate parameter changes and a flat cost function are well known (see, e.g., Daganzo, 1999). In our numerical experiments we have observed a similar behaviour for the RNM. Moderate changes in the system parameters result in small changes of the recovery network design, if any. For larger parameter variations the significance of network changes depends, in particular, on the investment costs for the disassembly centres. Sensitivity tends to increase along with investment costs until only one centre is opened. Other factors that tend to increase return network sensitivity include a decreasing minimum disposal fraction γ and, as for the forward network, a decreasing number of potential locations.

Given the structural parallels with traditional warehouse location models there does not seem to be a strong reason for requiring essentially new approaches to deal with uncertainty in recovery network models. In other words, while the general *level* of uncertainty can be expected to be significantly higher in a product recovery environment its *consequences* for the logistics network design do not seem to be more dramatic than in other contexts. Therefore, a deterministic MILP approach together with scenario analyses seems appropriate for recovery network design. While it is certainly true that stochastic or robust network optimisation may yield solutions that are not optimal for any single scenario, the above observations suggest that the cost difference with a solution from a detailed scenario optimisation is small. Moreover, the computational effort for evaluating a large number of scenarios seems more than counterbalanced by the exploding problem sizes of stochastic or robust optimisation models, unless special problem structure can be exploited.

Rather than for taking into account stochastic variations, the dependence of the network structure on the return volumes may be more important for long–term non–stationary considerations. Since product recovery is a fairly recent field many companies gradually extend their collection and recovery activities from moderate scale pilots to large scale business processes. For this strategic transition, multi–period extensions of the RNM may be worth considering. To keep problem sizes tractable exploiting the above results concerning the insensitivity of the forward network may be helpful. Results by Realff et al. (1999) provide a first step in this direction. Eskigun and Uzsoy (1998) have proposed another model that is worth mentioning in this context, considering product recovery management in a capacity extension setting. Furthermore, insights from Stuart et al. (1999) in a slightly different context may be valuable. Rather than considering geographical aspects, the authors address an optimal process design during a product lifecycle. For this purpose, a large–scale multi–period MILP production planning model has been developed, which explicitly takes into account dependencies between sales and future returns. Fixed costs are charged for intermediate process adjustments.

Table 5.4 summarises the above empirical results and lists the major factors determining the impact of product recovery on the logistics network design. We conclude this section by applying our observations to estimate network robustness for the set of case studies discussed in Chapter 4. See Table 4.1 to recall the major characteristics of each case. Figure 5.7 places each of these cases in a two–dimensional space indicating network sensitivity analogous with the above analysis. (The numbering refers to Section 4.1.) The horizontal and vertical axis refer to sensitivity with respect to return flows of the forward and return network, respectively.

We can divide the cases in two major clusters depending on the relation between disposer and reuse market. The first cluster contains cases with a closed loop structure and a close link between forward and reverse channel. The second group refers to cases where disposer market and reuse market are commercially and geographically separated. From our analysis we conclude that product recovery can be expected to be easy to integrate efficiently in existing logistics structures for the examples of the first cluster. This is in line with the real-life solutions we find in the cases. In 4 out of the 5 cases of this cluster the recovery network is built upon a previously existing logistics structure. Further differentiation within the groups is based on the factors discussed above. In contrast, the cases in the second cluster may require an integral network design approach according to our findings. However, since product recovery forms an entirely new business channel in all of these cases

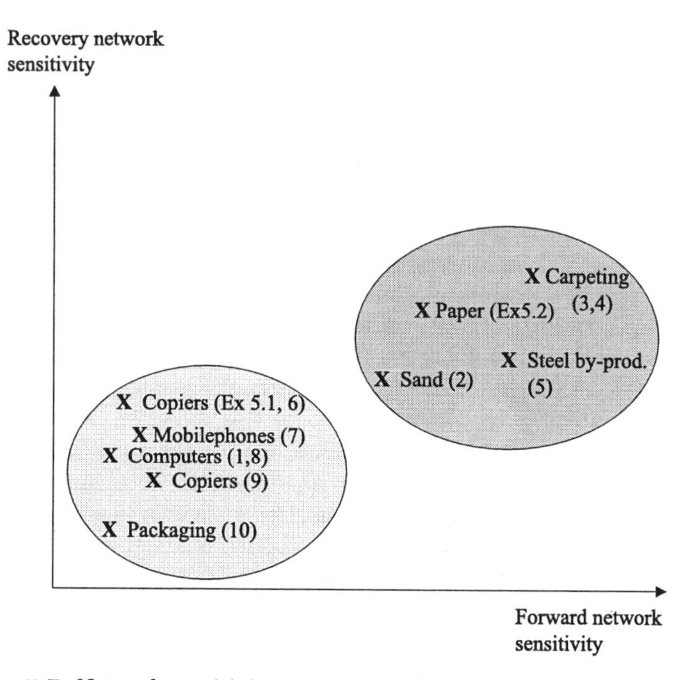

Fig. 5.7. Network sensitivity to return variations in exemplary cases

rather than supplementing existing 'virgin' production, a sequential network design does not seem a natural option anyway. Hence, the need to consider the entire network simultaneously appears not to be conflicting with the companies' business situations here.

5.5 Extensions

As pointed out before, the RNM is meant as a basic model capturing the major aspects of logistics network design in a product recovery context. The model can be extended in many ways to address more specific situations. In this section we discuss some extensions of the RNM which seem particularly relevant in a product recovery context. As discussed in Section 5.2, the model can be characterised as an uncapacitated, static, single–commodity, deterministic facility location problem. Analogous with the traditional location problem, the RNM can be modified into a capacitated, dynamic, multi–commodity, stochastic model. Moreover, it can include both revenues and costs as objective function and can be used in a multicriteria optimisation context. We do not consider these extensions in detail here since they are well known from other contexts (see, e.g., Mirchandani and Francis, 1989). Instead, we focus on additional elements that appear to be specific of product recovery, namely economics of integration and technology impact. We shortly discuss these issues below and show how to integrate them in the RNM formulation.

Integrating forward and reverse locations
Installing multiple facilities at the same location may give rise to synergies due to, e.g., sharing buildings and common equipment. This may be relevant, in particular, in considering the integration of forward and reverse channel activities. On the internal logistics level, including warehouse planning, integration versus separation of inbound and outbound flow handling has been mentioned as one of the major Reverse Logistics issues (Rogers and Tibben–Lembke, 1999; de Koster and van de Vendel, 1999). On the other hand, it should be noted that the forward and reverse network parts turned out to be quite similar in the above numerical examples, even without taking synergies into account explicitly.

In the model, synergies of integration may be captured by a reduction of fixed costs. We illustrate this approach for the case of combining warehouses and disassembly centres. Define:

f_l^s = savings in fixed costs for opening an integrated warehouse–disassembly facility at location l, as compared to $f_l^w + f_l^r$, $l \in J \cap L$.
Y_l^s = indicator opening an integrated warehouse–disassembly facility at location l, $l \in J \cap L$.

In the RNM add $-f_l^s Y_l^s$ to the objective function and add the following three constraints: $Y_l^s \leq Y_l^w$; $Y_l^s \leq Y_l^r$; $0 \leq Y_l^s \leq 1$. Note that in any optimal solution we have $Y_l^s = \min(Y_l^w; Y_l^r)$ and hence the new decision variable Y_l^s will automatically be integer valued.

Combining forward and reverse transportation
In a similar way, we can model joint distribution and collection. The benefit of combined transportation is clear when a warehouse and a disassembly centre are located together and the same fleet is used for delivery and collection. In the extreme case one may assume that collection has no due dates and can be carried out along with the next delivery visit, using the forwarding vehicles at no extra costs. Typically however, even if combined, collection activities do imply additional costs due to the use of additional resources (Beullens et al., 1999a,b). On the strategic decision level we suggest to model the savings of combined transportation on a given path as proportional to the minimum of the corresponding forward and reverse flow. Note that this is in line with current practice, e.g., in the softdrink industry where collection of reusable packaging is sometimes taken into account in the form of a cost supplement to 'forward' transportation costs. Define:

c_{kli}^s = unit variable cost savings for combining transportation on reverse path kli with transp. on forward path ilk, $l \in J \cap L$.
X_{kli}^s = fraction of total returns of customer k on path kli combined with deliveries on path ilk, $l \in J \cap L$.

For each path ilk for which potential savings are defined we now add the term $-c^s_{kli} r_k X^s_{kli}$ to the objective function. Moreover, we add the following three constraints: $r_k X^s_{kli} \leq d_k X^f_{ilk}$; $X^s_{kli} \leq X^r_{kli}$; $X^s_{kli} \geq 0$. Analogous with the previous example we have $r_k X^s_{kli} = \min(d_k X^f_{ilk}; r_k X^r_{kli})$ in any optimal solution. This approach may be further generalised by including savings from combined transportation of any two forward and reverse streams even if facilities and/or customers do not coincide.

Selecting recovery processing technologies

Different technologies may result in different processing costs and different recovery yields. Moreover, applicable technology may be volume dependent. For example, Arola et al. (1999) have compared an automated sorting line for plastics from disassembled electronic goods with manual sorting. Automated sorting turns out to be preferable only at high throughput rates. Since our network design involves decisions on the number of disassembly centres and the assigned processing volumes the selection of the best recovery technology and the outcomes of the model may be interdependent. In that case, one may want to integrate technology selection into the RNM.

To this end, we can follow the approach presented in the Multi–Activity Uncapacitated Facility Location Problem (Akinc, 1985). In addition to fixed costs for opening recovery centres we include technology specific fixed and variable costs associated with implementing and operating a specific technology at a specific recovery centre. For selecting the mix of processing technologies at each site that minimizes total costs define:

c^r_{klim} = unit variable cost of returns from customer k via l to i using technology m.

X^r_{klim} = fraction of returns from k via l to i using technology m.

f^r_{lm} = fixed cost to install processing technology m at disassembly centre l.

Y^r_{lm} = indicator installing processing technology m at disassembly centre l.

γ_m = minimum disposal fraction of technology m

In the objective function and in constraints (5.2), (5.3) and (5.4) we then substitute c^r_{kli} by c^r_{klim}, X^r_{kli} by $\sum_m X^r_{klim}$, X^r_{kl0} by $\sum_m X^r_{kl0m}$ and γ by γ_m. In (5.7) we replace X^r_{kli} by X^r_{klim} and Y^r_l by Y^r_{lm}. Moreover, we add the term $\sum_l \sum_m f^r_{lm} Y^r_{lm}$ to the objective function and introduce the following additional sets of constraints: $Y^r_{lm} \leq Y^r_l \quad \forall\ m, l$; $Y^r_{lm} \in \{0; 1\} \quad \forall\ m, l$; $0 \leq X^r_{klim} \leq 1 \quad \forall\ k, l, i, m$. Finally, we can relax the integrality constraints concerning the variable Y^r_l.

In addition to the above model extensions, it seems worthwhile to take a look at how certain policies may be used to influence parameter values. We briefly indicate three examples.

Value of information concerning quality of returns
Knowing product quality as soon as or even before products are returned by customers can result in a number of advantages. First of all, this allows for better maintenance during use and a better return policy depending on the product's life–cycle, which again may lead to a higher recovery potential (a lower value for γ) and lower recovery costs (lower c^r_{kli}). Second, inferior return products may be disposed of directly or treated locally, without shipping to a disassembly centre. Again, this results in a lower value of γ in the model. Knowledge on product quality can, for example, be supported by modern information technology including sensor–based data recording devices, electronic data logs and information systems for product recovery (Klausner et al., 1999).

Regional legislative requirements
We define $R = \{1, ..., N_e\}$ regions with different local regulation and let K_e indicate the set of customers in region e. One of the measures in legislative proposals concerns the amount of goods diverted from landfill. A minimum recovery level ρ_e as a percentage of total returns in region e can be incorporated in the RNM by adding the constraint $\sum_l \sum_{k \in K_e} X^r_{kl0} \leq (1 - \rho_e) \sum_{k \in K_e} r_k$.

End–of–life management
Enhancing the product recovery strategy may be a measure to change model parameters. Consider the following examples: (1) *product eco–design* could lead to different forward flow costs c^f_{ijk}, lower reverse flow costs c^r_{kli} and a higher recovery potential reflected in a lower value of γ; (2) a *buy–back scenario* where a cash payment is offered to customers for returning end–of–use products may result in higher average costs c^r_{kli} but also in a higher return rate r_k and, if refunding depends on the product quality, a higher recovery potential (compare Klausner et al., 1999); (3) *contract redesign* from sales towards lease contracts may lead to both a higher return rate r_k and a higher recovery potential, at the expense of higher forward flow costs.

Conclusions of Part II

The past two chapters have addressed distribution management in a Reverse Logistics context. In particular, logistics network design issues have been analysed. A review of ten recent case studies has illustrated that many companies are concerned with setting up logistics infrastructures for the recovery of used products. Moreover, the available business examples have been shown to be rather similar both in their scope and in the overall network structure. In all cases the recoverer considers goods flows beginning with the collection of used products and ending with the distribution of recovered products. Intermediate activities include inspection and separation, re–processing, and disposal steps. The typical logistics network structure encompasses a convergent collection part, a divergent distribution part, and an intermediate part related with the specific recovery processing steps. In particular, these networks encompass both 'reverse' and 'forward' flows and hence show Reverse Logistics to be a subset of a company's overall logistics task.

A lack of control on the supply side both with respect to quantity and quality appears to be the major distinction between product recovery networks and traditional production-distribution networks. While in traditional supply chains supply is selected as a function of demand, Reverse Logistics inbound flows are partly exogenously determined. Rather than being entirely demand driven product recovery networks therefore involve supply push drivers. As a consequence, companies face a significant level of supply uncertainty on the one hand and the need to balance supply and demand on the other hand. Considering recovery environments in more detail, including product, market, and resource aspects we have seen that product recovery networks can be subdivided into a number of classes. In particular, the form of recovery appears to be a discriminating factor. Re-usable item networks, remanufacturing networks, and recycling networks have been shown to have their own typical characteristics.

Based on the insights from the case study analysis we have proposed a MILP model to support recovery network design. The model is similar to a multi–level warehouse location model. One main difference concerns additional constraints, reflecting the need for coordination between supply and demand. To some extent, this effect can be interpreted as a capacity constraint. Furthermore, the model involves an additional degree of freedom due to returns dispositioning. The model has been shown to be fairly general and to encompass different recovery situations, including closed loop versus open loop structures, push versus pull drivers, and the possible admission of a regular production source alternative to product recovery.

The model has been illustrated numerically in two cases concerning copier remanufacturing and paper recycling. Moreover, this analysis has allowed to investigate the impact of Reverse Logistics flows on the layout of logistics networks. In particular, we have addressed the question of whether adding a recovery network to an existing forward network entails substantially higher costs than the simultaneous design of both the forward and the reverse network part.

Numerical results have shown product recovery networks to be fairly robust in several respects. As in many traditional distribution networks, moderate parameter changes result in small changes, if any, in the optimal facility locations and the corresponding network design. What is more, we have seen that forward flows, in general, dominate the network layout. Return flows appear to have a significant impact on the overall network structure only in the case of both a major structural difference between forward and reverse channel costs and high return volumes. The latter case has been illustrated in the paper industry example where recycling reduces the dependence on the raw material sources, which largely dominated the original forward channel structure. Consequently, product recovery 'pulls' activities geographically closer to the customers. However, in many other cases, such as in the copier remanufacturing example, Reverse Logistics flows do not have a relevant impact on the structure of the outbound network part.

This is good news since product recovery can in many cases be implemented without requiring major changes in existing 'forward' production–distribution networks. Moreover, separate networks can be expected to be easier to deal with organisationally. A company can create a new, dedicated organisational unit to deal with return flows. Therefore the cost of coordination and restructuring tends to be lower.

From a methodological point of view the observed robustness means that forward and return networks can be modelled separately in many cases, which significantly reduces the problem sizes. Moreover, the experimental results lead us to the conclusion that supply uncertainty has a limited effect on the logistics network design and that deterministic modelling approaches appear to be appropriate for recovery network design in many cases. Long-term non-stationary effects such as starting up and extending product recovery activities may be an argument for multi–period models, which certainly deserve further attention.

Part III

Reverse Logistics: Inventory Management Issues

6. Inventory Systems with Reverse Logistics

As explained in Chapter 1 this part of the book is dedicated to inventory management issues in Reverse Logistics. Complementing the spatial considerations addressed in the preceding chapters inventory management is concerned with the temporal co–ordination of subsequent business processes and, eventually, between supply and demand. Inventory management has been investigated in much detail during the past fifty years. Different reasons for stock–keeping have been distinguished, such as lotsizing stock, safety stock, and seasonal stock. Corresponding management issues that have been addressed include determination of process decoupling points, optimal order policies, and expected stock levels. Countless quantitative models have been proposed for inventory management including, in particular, numerous classes of deterministic and stochastic inventory control models. We refer to Silver et al. (1998) for a detailed discussion of inventory management. Recent developments in supply chain management have led to an increasing effort to reduce inventory levels on a global scale. Just–in–time philosophies and vendor–managed–inventories are some of the concepts pointing in this direction (compare Tayur et al., 1998).

It will come as no surprise that inventory management also plays a role in a Reverse Logistics context. As discussed in Section 4.3 inventories may be kept between any two subsequent activities of a product recovery network. For example, one may postpone transportation of collected used products to an inspection site until a sufficient number of products has accumulated. Similarly, inspected products may only be reprocessed once demand has depleted previous stocks. One may think of similar examples for other Reverse Logistics channels, e.g., reuse of commercial returns or impairment and disposal of sensitive asset returns (compare Table 1.1).

In this chapter we investigate inventory management issues in a Reverse Logistics context in more detail. To this end, we take a look at a number of case studies in Section 6.1 below. Although material available from literature turns out to be fairly limited we observe some common characteristics in Section 6.2 and compare them with a conventional supply chain context. Section 6.3 concludes this chapter with a detailed literature review on mathematical inventory models in Reverse Logistics.

6.1 Exemplary Business Cases

Probably the best known example of Reverse Logistics related inventory management concerns rotable spare parts. In many applications, both civilian and military, spare parts for maintenance are kept in a closed loop as much as possible. Upon failure (or preventive maintenance) a part in the field is replaced by a spare part from inventory. Subsequently, the failed part is returned to the maintenance provider, inspected and repaired if possible and added to the spare parts inventory again. Examples are manifold and include military equipment, aircraft and railway engines and machine tools. Moreover, we recall from Chapter 2 the example of IBM's spare parts network for computer components. Inventory control in rotable spare parts systems has been investigated extensively since the 1960s. Sherbroke's METRIC model introduced in 1968 has become a standard, which is at the basis of many subsequent contributions. Rotable spare parts systems are characterised by a closed–loop behaviour where the number of parts is constant. Consequently, demand and returns are highly correlated. Every return of a failed part is accompanied by a simultaneous demand for a replacement part. Conversely, since not all returned parts may indeed be repairable new parts may have to be injected into the system from time to time. Major issues in rotable spares inventory management include determining the number of parts in the system, the number and location of stock points, and the allocation of inventories.

Literature on business practice concerning other examples of inventory management in a Reverse Logistics context is surprisingly scarce. While it is not difficult to imagine corresponding issues, as sketched in the introduction to this chapter, detailed case descriptions are widely lacking. In what follows we discuss a number of sources providing initial information in this direction.

- In a recent study, Elmendorp (1998) has addressed the management of reusable packaging by a major brewery in Belgium and The Netherlands. The company is using a deposit based system encompassing two types of bottles, some twenty types of crates, and four types of barrels covering the major share of products sold on the national market. Products are distributed via retailers and a company–owned trade organisation. Upon delivery of new products empty packaging is collected from the retailers and the trade organisation back to the brewery where it is stored, cleaned and refilled. Availability of empty packaging is an important constraint on the production process. The study distinguishes three planning levels concerning the inventory of reusable packaging material. On a long–term basis the total numbers of bottles, crates, and barrels required are assessed, based on sales forecasts and estimated circulation frequencies. Seasonality is an important factor at this stage. On a medium–term basis actual orders for new packaging are placed. On a short–term basis bottles may be placed in different crates, the production sequence may be adjusted, and taking back packaging from retailers may, to some extent, be expedited. The study reports in detail on the forecasting of packaging returns. Based on the data available at the company, namely ingoing and outgoing flows at the brewery, it turns out that delivery and returns develop largely in parallel, as a consequence of the companies collection policy. Given this absence of a visible time lag, the author concludes that returns may be modelled as an autonomous time series and that including sales information does not improve forecasting quality.

- Toktay et al. (1999) consider inventory management for Kodak's single–use cameras (see also Chapter 1). Printed circuit boards for the production of these cameras are either bought from overseas suppliers or remanufactured from the cameras returned by the customers via photo laboratories. The issue is to determine a cost–efficient order policy for the external supplies. Major difficulties arise from the fact that return probabilities and market sojourn–time distribution are largely unknown and difficult to observe. The authors propose a closed queueing network model to address these issues. They consider a base stock order policy and develop a Bayesian approach to update estimates of the unknown parameters. The value of more detailed information concerning the return process is then assessed by comparing the long–run average costs of different informational structures, including ignoring past sales; estimating the return distribution based on aggregate data and based on individual camera data; and complete information on the return distribution. The authors conclude that estimating the number of cameras in the market based on sales information can reduce inventory costs significantly. Moreover, estimating the distribution of the market sojourn–time on an aggregate basis yields satisfactory results such that tracking return data per camera does not seem to pay off. The latter confirms earlier results by Kelle and Silver (1989a) in the context of reusable containers.
- A somewhat surprising application of inventory control with Reverse Logistics is pointed out by Rudi and Pyke (1999). They report on a study at the Norwegian National Insurance Administration concerning the supply of medical devices such as wheelchairs and hearing aids to handicapped people. In the cadre of the Norwegian health insurance system people in need are supplied with these medical devices free of charge. Once the devices are no longer needed they are returned to the insurance administration where they are inspected. Based on their state they are then either refurbished and stocked for future use or scrapped, possibly after dismantling reusable parts. The issue in this system is to control the purchasing of new devices and to decide whether or not to refurbish a returned item. The authors present a simple support system for the refurbishment decision assuming relevant costs and benefits to depend linearly on the age of a specific item.

Recently, several studies have addressed the use of end–of–use returns as an alternative source for spare parts. Integrating used product returns extends the classical spare parts systems discussed above. In particular, the closed loop character of the system is relaxed since there is no direct link, in general, between used product returns and demand for spare parts.

- As an example consider the case of IBM presented in Chapter 2. As discussed, used computer equipment which cannot be refurbished may be dismantled to recover reusable spare parts. Dismantling is cheaper both than purchasing new parts and than repairing failed parts from the field in many cases. On the other hand, availability of used products for dismantling turns out to be difficult to forecast, which complicates order decisions on other sources. Note in particular that used product returns, which may be supposed to depend primarily on past sales or lease issues, are not related with demand for spare parts and may therefore raise parts inventory levels. We address the integration of dismantling into IBM's spare parts planning in detail in Chapter 9.
- Driesch et al. (1997) report on a similar situation for car parts at Daimler–Benz. Broken engines returned by customers are remanufactured in a dedicated recovery facility for use either as entire engine or as collection of spare parts. In both cases engines are disassembled completely, which gives rise to two kinds of component inventories, namely dismantled parts of yet unknown quality and remanufactured, i.e. reusable, parts. Again there is no direct one–to–one relation between inbound and outbound goods flows since a customer returning a failed engine does not necessarily purchase a remanufactured engine and has no link with spare parts demand either. The authors propose an MRP–scheme for controlling the purchasing and recovery operations in this system.
- Finally, Van der Laan (1997) investigates a similar system at Volkswagen. The company offers remanufactured car exchange parts as a cheaper alternative to new parts. For this purpose, used car parts are taken back through national importer organisations to a central recovery facility in Germany. Returned parts are remanufactured by a subcontractor and are then sold by Volkswagen through the national importers to car dealers. If remanufactured parts do not cover demand new parts are offered for the same price. The challenge in this system is to control the inventories of returned and remanufactured parts. Note that even upon looking at spare parts only, this system still differs from classical examples in that demand and supply are not directly linked. In the case study a simple EOQ–based reorder policy for the remanufactured inventory is investigated. Performance is found to be satisfactory if parameters are updated regularly, corresponding to the product life–cycle.

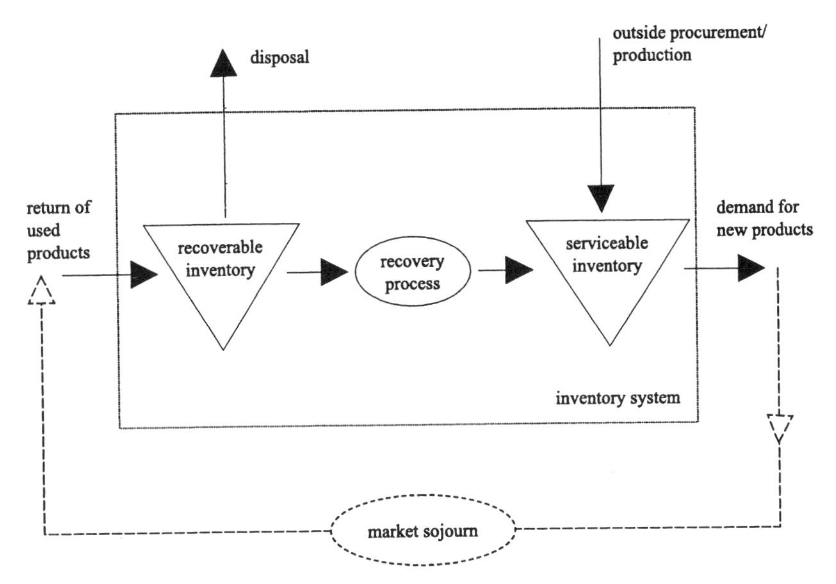

Fig. 6.1. Framework for inventory control with Reverse Logistics

6.2 Characteristics of Recoverable Inventory Management

Although the material on current business practice concerning inventory management in a Reverse Logistics context is rather limited a number of observations can be made, considering the examples in the previous section. From an inventory control perspective, all the cases discussed have a fairly similar structure. Figure 6.1 presents a general framework, which is adapted from Van der Laan (1997).

In all of the examples one may distinguish items in two states namely recoverable items returned from a market and serviceable items that can directly meet demand. Some recovery process transforms recoverable items into serviceables. As examples consider empty versus re–filled beverage containers (see Elmendorp, 1997), returned versus inspected printed circuit boards (see Toktay et al., 1999), and used versus remanufactured car parts (Van der Laan, 1997). In general, both recoverable and serviceable items may be stocked, leading to a system of two serial inventories.

The supply of returned items is fully exogenously determined in all of the above cases. The recoverer has no means to influence the timing or quantity of returns. Instead, returns forecasting appears to be a major issue. Moreover, excess stock may possibly be disposed of. In addition, all examples include an alternative supply source that is fully controllable to replenish the serviceable stock. One may think of the supply of 'virgin' printed circuit boards (Toktay et al., 1999), medical devices (Rudi and Pyke, 1999), and computer spare parts (see Chapter 2) for example.

It should be noted that the above situation resembles the distribution management setting discussed in Section 4.3 in that the logistics system considered forms a link between two exogenous drivers, namely (recoverable) supply and demand. The relation between both processes appears to be one main discriminating factor characterising different recovery situations. As discussed above, classical spare parts systems rely on a closed loop situation where every return triggers an instantaneous demand. In other situations the causal relation appears to be reversed, in that demand entails a subsequent return, typically after a certain time lag. We recall the cases of single–use cameras (Toktay et al., 1999) and beverage containers (Elmendorp, 1997) as examples. In yet other situations, there may be no closed loop at all and no direct relation between demand and returns. As discussed above the latter holds, for example, for the returns of used computer equipment for dismantling and the demand for spare parts at IBM (see Chapter 2).

In addition, note that the above framework once more encompasses both Reverse Logistics flows and conventional 'forward' flows, just as the location models discussed earlier. In this sense, the above cases provide another argument for not considering Reverse Logistics in isolation but embedded in the overall logistics context.

Comparing Figure 6.1 with a traditional inventory control context (see, e.g., Silver et al., 1998) two major distinguishing factors can be observed, namely an exogenous inbound flow on the one hand and multiple supply options for the serviceable stock on the other. However, both features as such are not new. Parallels and differences with classical spare parts systems have already been discussed above. As pointed out by Van der Laan (1997) another inventory control context with exogenous inbound flows is cash balancing. This refers to the management of a bank's local cash which is raised by customer deposits and lowered by customer withdrawals. The cash level may be controlled by transfers to and from a central cash. Cash balancing models have been studied in literature since a long while (see Inderfurth, 1982, for a review). Considering the above framework they may provide a potential starting point for developing inventory models in a Reverse Logistics context. While we defer the detailed analysis of corresponding quantitative models to subsequent chapters we note at this point that application of cash balancing models in a Reverse Logistics context is mainly limited by its restrictive leadtime assumptions. Transfers between local and central cash usually do not consume much time. In contrast, leadtimes for both the recovery process and 'virgin' supply in a Reverse Logistics context may be substantial and, in addition, different for both channels.

Previous inventory models with multiple supply options mainly concern emergency supplies (see, e.g., Moinzadeh and Nahmias, 1988). Typically, a slow but cheaper supplier and a fast but more expensive supplier are considered. Orders are placed at the cheap supplier unless the inventory level drops below a critical point. In that case, an emergency order is placed at the fast supplier to avoid stockouts. In this way, emergency supply models are a direct extension of lost–sales models. The major difference between this setting and the framework as displayed in Figure 6.1 is the fact that both the regular and the emergency supplier are always available. In contrast, availability of product returns for recovery is exogenously determined. Hence, there may not always be a choice as to which source to use at a given moment. Nevertheless, emergency supply models may be worth considering as a starting point for Reverse Logistics inventory models. We analyse the impact of both the exogenous inbound flow and the alternative supply options on quantitative inventory models in detail in Chapters 7 and 8.

6.3 A Review of Inventory Models in Reverse Logistics

Given the limited number of published case studies on inventory management in a Reverse Logistics context it is remarkable that a fairly substantial number of quantitative models has been proposed for this issue. Some are developed as extensions to classical spare parts models. Others are inspired by the recent interest in Reverse Logistics and product recovery management. In general, the contributions tend to be mainly mathematically oriented and the link with business examples is rather loose. In this section we review literature concerning quantitative inventory control models within the scope of the framework sketched in Figure 6.1. From a mathematical inventory theory perspective, deterministic and stochastic models can be distinguished, and the latter can be further subdivided into periodic and continuous review models. We treat each of these groups separately below.

As pointed out before, repairable spare parts models have been investigated in much detail for at least four decades. Revisiting this material does not seem to add to our analysis. Therefore, we do not consider models in the sequel where returns are limited to product or parts failures that immediately generate a replacement demand. For a detailed discussion of repairable spare parts models we refer to the standard reviews by Pierskalla and Voelker (1976), Nahmias (1981), and Cho and Parlar (1991). Additional recent references are discussed by Guide and Srivastava (1997).

6.3.1 Deterministic Models

In deterministic inventory control models information on all the components of the framework presented in Figure 6.1 is assumed to be known with certainty. In particular, demands and returns are known a priori for the entire planning horizon. Using the taxonomy of inventory theory Table 6.1 lists deterministic models from literature that fall within the scope of the above

Table 6.1. Deterministic inventory models with Reverse Logistics

	Schrady (1967)	Mabini et al (1992)	Richter (1994)	Richter (1996,1997)	Teunter (1998)	Richter and Sombrutzki (2000)	Beltran and Krass (1997)	Minner and Kleber (1999)
cont./discr. demand	c	c	c	c	c	d	d	c
cont./discr. actions	d	d	d	d	d	d	d	c
dynamic / stationary	s	s	s	s	s	d	d	d
planning horizon	∞	∞	∞	∞	∞	T	T	T
cost criterion	avg	avg	avg	avg	avg	tot	tot	tot
# stockpoints	2	2	2	2	2	1	1	2
# decision variables	3	3	2	4	3	1	2	3
disposal	-	-	+	+	+	-	+	+
fixed costs	+	+	+	+	+	+	+	-
leadtimes	+	+	-	-	-	-	-	-
optimisation	par	par	par	par	par	pol	pol	pol

framework. For each model discrete versus continuous demand and decision epochs are specified as well as a stationary versus dynamic approach. Moreover, the planning horizon and the objective function's cost criterion are indicated. Some of the models explicitly take into account the two types of inventory distinguished in Figure 6.1 whereas others consider a single aggregated stock–point only. Similarly, a varying number of decision variables may be involved. In particular, disposal of excess returns may or may not be allowed. In addition, we show whether fixed costs and leadtimes are included. Finally, we distinguish research contributions focussing on the cost evaluation of a fixed decision policy class, parametric optimisation within a given class of policies, and global optimisation of the policy structure.

The deterministic models can primarily be subdivided into static and dynamic models. The former correspond with the mindset of the classical economic order quantity (EOQ) seeking an optimal tradeoff between fixed setup and variable holding costs. Several authors have proposed extensions to this model taking return flows into account. A first model of this type was proposed by Schrady as early as in 1967. The model comes down to the system in Figure 6.1 with constant demand and return rates and fixed leadtimes for external orders and recovery. Disposal is not allowed. The costs considered are fixed setup costs for external orders and recovery and linear holding costs for serviceable and recoverable inventory. The author proposes a control policy with fixed lotsizes for procurement and recovery where each procurement order is followed by n identical recovery batches. Expressions for the optimal value of n and for the optimal lotsizes are derived analogous with the classical EOQ model. In particular, the optimal procurement–lotsize equals the EOQ formula for the 'net' demand rate (i.e. demand minus returns) in the case of identical holding costs for serviceables and recoverables.

Mabini et al. (1992) have proposed an extension to the above model including a stockout service level constraint. Moreover, a corresponding multi–item system is discussed in which different items share the same repair facility. Numerical solution methods are proposed for both situations.

Subsequently, Richter (1994) has considered Schrady's model for alternating procurement and recovery batches (i.e. $n = 1$ in the above setting) and analyses the dependence of the cost function on the return rate. He shows that costs are convex in the return rate if holding costs for recoverables do not exceed serviceable holding costs. This result may be useful if the return rate can be influenced, e.g., by means of disposal. Richter (1996, 1997) extends the analysis to the case of multiple consecutive procurement and recovery batches. He shows that the optimal solution to a relaxed problem, where the number of setups may be non–integer, uses multiple consecutive batches for at most one of the sources. Moreover, this result is used to characterise situations where it is advantageous to achieve return rates of 0% or 100%, respectively.

Teunter (1998) considers the same model anew for a modified disposal policy. Rather than assuming a constant disposal rate all returns occurring during a certain time span are disposed, while all returns thereafter are accepted again. Disposal involves a linear cost per item. Moreover, the author assumes different holding costs for recoverable, recovered, and procured items. Expressions for the optimal lotsizes in this policy are derived. Furthermore, it is shown, as in Richter's analysis (1996), that the number of consecutive batches equals one either for recovery or for procurement (or both) in any optimal solution when ignoring integrality constraints on the number of setups.

It should be noted that all of the above results are somewhat heuristic in that they optimise parameter values for predetermined policies without studying optimality of the policy itself. To the best of our knowledge there are no results that specify the structure of an optimal policy.

Besides the above static models a few dynamic lotsizing models similar to the classical Wagner–Whitin–model (Wagner and Whitin, 1958) have been proposed for a Reverse Logistics context. Most of these models consider a single stock point, hence aggregating the two types of inventory distinguished in Figure 6.1. Richter and Sombrutzki (2000) discuss applicability of the original Wagner–Whitin–model in Reverse Logistics situations. Reversing the time axis they argue that the traditional model can be interpreted as looking for optimal recovery batches for returned products that are accumulating. Note that this interpretation does not include an alternative procurement option nor disposal. Moreover, demand is not a restricting factor.

Beltran and Krass (1997) consider dynamic lotsizing for an inventory point facing both demand and returns. This comes down to the original Wagner–Whitin–model with the exception that (net) demand may be positive or negative. Moreover, the inventory may be both raised by procurement and decreased by disposal. Both actions involve fixed plus concave costs. The authors show that the zero–inventory–property, which is well known for the original model, has to be modified here. Rather than assuming zero inventory at any order epoch, it can be shown that there is a period of zero inventory between any two actions (procurement or disposal) in an optimal policy. Due to returns a procurement order may sometimes be delayed beyond the first occurrence of inventory depletion. This more general structure complicates the computation of an optimal policy. The authors propose a dynamic programming algorithm which is of complexity $O(N^3)$ in the general case and of $O(N^2)$ under additional restrictions on the cost functions.

Finally, Minner and Kleber (1999) address the situation in Figure 6.1 in an optimal control context. In addition to demand and returns, all actions, namely procurement, recovery, and disposal are modelled as non–stationary continuous processes. Optimality conditions are derived under a linear cost structure. The optimal control policy is shown to change at discrete points only and be extremal, in the sense that recovery and disposal actions always cover all or nothing. This allows for an easy solution algorithm. Results are illustrated in a scenario with seasonality and a fixed time lag between demand and returns.

6.3.2 Stochastic Periodic Review Models

As listed in Table 6.2 several periodic review inventory models have been proposed in literature that fit in the framework of Figure 6.1. In addition to the characteristics discussed in the previous subsection the form of stockout handling is indicated for each model, namely backordering versus lost sales. Moreover, we specify assumptions on the demand and return processes and on their relation. As discussed in Section 6.1 both processes may be linked by a fixed or stochastic time lag or be independent.

In most of these contributions an optimal policy is sought for procurement, recovery, and/or disposal decisions. One main distinction can be made between models considering one aggregated stock point and more detailed models explicitly taking into account the two types of inventory indicated in Figure 6.1. Within the former class, models differ mainly with respect to the assumptions concerning the relation between demand and returns. Whisler (1967) analyses a single inventory where each issued item returns after a stochastic market sojourn time. Similar to Beltran and Krass (1997,

Table 6.2. Periodic review inventory models with Reverse Logistics

	Whisler (1967)	Kelle and Silver (1989)	Toktay et al (1999)	Buchanan and Abad (1998)	Cohen et al (1980)	Beltran et al (1997)	Simpson (1970)	Simpson (1978)	Inderfurth (1996,1997)	Inderfurth et al (1998)
dyn. / station.	d	s	s	s	s	d	s	s	s	s
planning horizon	T/∞	T	∞	T	T	T/∞	∞	T	T	T/∞
cost criterion	disc	tot	avg	tot	disc	tot/disc	avg	disc	tot	disc/avg
stockout	lost	back	lost	back	lost	lost	back	back	back	back
# stockpoints	1	1	1	1	1	1	1	2	1 / 2	$N+1$
# items	1	1	1	1	1	1	1	1	1	1
# decision vars.	1	1	1	1	1	1	1	3	2 / 3	$N+1$
disposal	+	-	-	-	-	-	-	+	+	+
fixed costs	-	+	-	-	-	+	-	-	-	-
leadtimes	-	-	+	-	-	-	+	-	+	+
demand process	gen	gen	Pois	gen	gen	gen	gen	gen	gen	gen
return process	gen	gen	gen	gen	gen	gen	gen	gen	gen	gen
relation dem. / returns	stoch lag	stoch lag	stoch lag	expon lag	fixed lag	lag 1	indep	indep	indep	indep
optimisation	pol	pol	par	pol	pol	pol	par	pol	pol	pol

see above), the inventory level may be increased or decreased instantaneously by means of procurement and disposal. Considering an equivalent queueing model the structure of an optimal control policy under linear costs is shown to be characterised by two critical numbers $L < U$. Whenever the inventory level at a review epoch lies outside the interval $[L, U]$ it is optimal to order up to L or dispose down to U, respectively. For intermediate inventory levels the optimal action depends on additional parameters. In the case of Poisson distributed demand and exponentially distributed market sojourn times it is optimal not to take any action.

A similar situation has been analysed by Kelle and Silver (1989). They assume issued items to be returned after a stochastic time lag or to be lost eventually. Since the average net demand is positive no disposal option is included. On the other hand, fixed procurement costs are taken into account. The authors formulate a chance–constrained integer program, which can be transformed into a dynamic lotsizing model with possibly negative demand, based on an approximation of the cumulative net demand. Note that this model differs from the extended dynamic lotsizing model discussed in Subsection 6.2.1 (Beltran and Krass, 1997) in that disposal is not included. Therefore, it can be transformed further into an equivalent conventional Wagner–Whitin model.

In their model concerning the case study on reusable cameras (see Section 6.1) Toktay et al. (1999) also assume a stochastic sojourn time and a certain loss fraction in the market. Based on a six node closed queueing network they determine an optimal base stock level to control procurement decisions assuming Poisson demand. As discussed above, much attention is paid to an efficient estimation of unknown parameters of the return process.

Buchanan and Abad (1998) modify Kelle and Silver's model by assuming for each period that returns are a stochastic fraction of the number of items in the market. Note that this comes down to an exponentially distributed market sojourn time. Moreover, in each period a fixed fraction of items from the market is lost. Under these conditions the authors derive an optimal procurement policy depending on two state variables, namely the on–hand inventory and the number of items in the market.

Cohen et al. (1980) consider a similar system assuming a fixed market sojourn time. Moreover, a given fraction of demand in each period will not be returned. In addition, a certain fraction of on hand inventory is lost due to decay in every period. The authors propose a heuristic order–up–to policy which is shown to be optimal in the case of a market sojourn time of one period. In the general case the target inventory level may occasionally not be attainable if returns are too high. The special case of a one period market sojourn time is also analysed by Beltran et al. (1997). Taking additional fixed costs into account they show optimality of a conventional (s, S)–reorder policy under some technical assumptions.

Finally, Simpson(1970) assumes demand and returns to be independent with a positive expected net demand. He proposes a heuristic for computing an order–up–to level under linear costs and a stochastic procurement leadtime when neglecting intermediate backorders cleared by returns.

A first model explicitly considering the two distinct inventories for serviceables and recoverables (see Figure 6.1) has been proposed by Simpson (1978). He considers the trade–off between material savings due to reuse of old products versus additional inventory carrying costs. Period demand and returns are modelled as generally distributed random variables that are independent except for possible correlation within the same period. Optimality of a three parameter (L, M, U) policy to control procurement, recovery, and disposal is shown when neither fixed costs nor leadtimes are involved. The policy can be interpreted as 'recover while serviceable stock is below M' and then adjust the echelon stock (i.e. the sum of both inventories) according to Whisler's (L, U)–policy.

Inderfurth (1996,1997) has extended Simpson's model by considering the impact of non–zero leadtimes both for procurement and recovery. He shows that it is the difference between both leadtimes which determines the system's complexity. If leadtimes are equal Simpson's policy can be shown to remain optimal by considering an appropriate inventory position rather than the net stock (analogous with Veinott's classical state reduction technique; see Veinott, 1966). In all other cases growing dimensionality of the underlying Markov model prohibits simple optimal control rules. A similar result holds if recoverables cannot be stored but need to be recovered or disposed of immediately. In this case Whisler's (L, U)–policy is optimal for equal leadtimes and for a procurement leadtime excess of one period. All other cases result again in fairly intractable situations.

Finally, Inderfurth et al. (1998) present a model for assigning returned products to alternative recovery options or disposal, which gives rise to a two–level divergent inventory system. Demand for each option and returns are again assumed to be independent. The expected total net supply from returns is positive and there is no additional procurement source in this model. The issue is to find an optimal policy for allocating returns to the alternative options taking into account holding and backorder costs. The authors show the optimal policy to have a complicated structure and propose a simple critical number policy as an approximation.

6.3.3 Stochastic Continuous Review Models

Finally, a number of continuous review models within the above framework has been proposed as listed in Table 6.3. All of these models are stationary and analyse the infinite horizon system behaviour. Focus is on a general cost analysis and on determining optimal parameter values for predetermined control policies. In contrast, results on optimal policy structures are few. In almost all cases demand and returns are modelled as independent Poisson process.

As in the previous subsection the proposed models can be divided into two groups considering, respectively, a single aggregated stockpoint or distinct recoverable and serviceable inventories. Within the former class Heyman (1977) analyses disposal policies to optimise the trade–off between additional inventory holding costs and production cost savings. He models demand and returns as general independent compound renewal processes. Since both recovery and procurement are instantaneous no stockouts occur. Ignoring fixed costs the system is controlled by a single parameter disposal level strategy: incoming returns exceeding this level are disposed of. The author shows equivalence of this model with a single server queuing model. For the case of Poisson distributed demands and returns he derives an explicit expression for the optimal disposal level. Furthermore, he proves optimality of the one parameter policy in this case. For generally distributed demands and returns an approximation is given.

Muckstadt and Isaac (1981) consider a similar model where the recovery process is explicitly modelled as a multi–server queue. In contrast with the above approach disposal decisions are not taken into account. The costs

Table 6.3. Continuous review inventory models with Reverse Logistics

	Heyman (1977)	Muckstadt and Iscaac (1981)	van der Laan et al (1996a,b)	Yuan and Cheung (1998)	Teunter (1999)	van der Laan et al (1999a,b)	van der Laan an Salomon (1997), Inderfurth and van der Laan (1998)
dynamic / stationary	s	s	s	s	s	s	s
planning horizon	∞	∞	∞	∞	∞	∞	∞
cost criterion	avg/disc	avg	avg	avg	NPV	avg	avg
stockout	–	back	back	back	–	back	back
# stockpoints	1	1	1	1	2	2	2
# items	1	1	1	1	1	1	1
# decision variables	1	2	3 / 4	2	3	3 / 4	4 / 5
disposal	+	-	+	-	-	-	+
fixed costs	-	+	+	+	+	+	+
leadtimes	-	+	+	-	-	+	+
demand process	comp.ren.	Pois	Pois	Pois	Pois	Pois	Pois
return process	comp.ren.	Pois	Pois	Pois	Pois	Pois	Pois
relation demand / returns	indep	indep	indep	exp.lag	indep	indep	indep
optimisation	pol	par	par	par	par	eval	eval

considered comprise serviceable holding costs, backorder costs, and fixed procurement costs. A control policy is proposed that controls procurement according to a traditional (s, Q)–rule whereas returned products directly enter the recovery queue. Values for s and Q are determined based on an approximation of the distribution of the net inventory. In a second step these results are carried over to a two echelon model.

Van der Laan et al. (1996a,b) propose an alternative procedure for determining the control parameters in the above (s, Q)–model based on an approximation of the distribution of the net demand during the procurement leadtime. A numerical comparison shows this approach to be more accurate in many cases. Moreover, the model is extended with a disposal option, for which several policies are compared numerically. The authors recommend to base disposal decisions on two critical numbers limiting the recovery queue length and an appropriately defined inventory position, respectively.

Yuan and Cheung (1998) model dependent demand and returns by assuming an exponentially distributed market sojourn time. Moreover, items may eventually be lost with a certain probability. Leadtimes for both recovery and procurement are zero and there is no disposal option. The authors propose an (s, S) reorder–order–up–to policy for procurement based on the sum of items on hand and in the market. It should be noted that this implies the number of items in the market to be observable. The long–run average costs for this policy are calculated based on a two–dimensional Markov process. A numerical search algorithm is proposed for finding optimal control parameter values.

Teunter (1999) distinguishes serviceable and recoverable stock and once more assumes demand and returns to be independent Poisson processes. The performance of an EOQ–based heuristic is evaluated. To this end, lotsizes for procurement and recovery are determined in a deterministic model (see Teunter, 1998, discussed above). Whenever the serviceable stock is depleted a recovery order is placed if enough recoverable items are on hand for the predetermined lotsize. Otherwise a procurement order is placed. Numerical results are given that document a good performance of this heuristic.

Van der Laan et al. (1999a,b) present a detailed analysis of different policies to control serviceable and recoverable stock in the above setting, taking into account non–zero leadtimes for both sources. In particular, a push- and a pull–driven recovery policy are considered. In the first case, all returned items are recovered as soon as the recoverable stock is sufficient to achieve a certain lotsize. In the second case, recovery is controlled by an (s, S) policy based on the serviceable inventory position, defined as serviceable inventory on hand minus backorders plus outstanding (recovery or procurement) orders. Procurement is controlled by an (s, S)–policy concerning the serviceable inventory position in both cases. Long–run expected costs for both policies are computed by evaluating a two–dimensional Markov process. Control parameter values are determined via enumeration. The authors indicate that the above inventory position has some drawbacks in the case of a large difference between the leadtimes of the two sources. Therefore, Inderfurth and van der Laan (1998) propose a modified inventory position for this case. The essence of this approach is to take only those outstanding orders into account that are within a certain time window.

Van der Laan and Salomon (1997) extend the above model to include a disposal option. For the pull-strategy an upper bound on the recoverable inventory is proposed as disposal trigger. For the push–strategy the recoverable inventory is limited by the recovery lotsize anyway and disposal is therefore controlled on the basis of the aforementioned inventory position. The authors show that a disposal option can significantly reduce the system costs by avoiding excessive stock levels, in particular for large return rates. A detailed numerical analysis is presented concerning the cost impact of various parameters.

7. Impact of Inbound Flows

As revealed in the previous chapter, one of the major distinguishing characteristics of inventory control in a Reverse Logistics context is the need for integrating a largely exogenously determined goods inflow. In this chapter we address this issue in more detail and quantify the impact of inbound goods flows on inventory dynamics and appropriate control strategies. For the sake of focus we start from a basic situation allowing for a detailed analysis. Subsequently, potential extensions and limitations of this approach are discussed. As with many of the models discussed in the previous chapter, we therefore aggregate the two inventory types distinguished in the framework of Figure 6.1 into a single stock point. Moreover, we assume demand and returns to be independent. Finally, we do not include a disposal option, hence procurement is the only means to control the system. One may view this setup as the common core of the models discussed in Chapter 6. Our goal is to identify appropriate decision rules in the above setting and to investigate the impact of the return flow on the system's performance.

The material of this chapter is organised as follows. In Section 7.1 we formally introduce our model and notation. Section 7.2 addresses the special case of unit demand and return quantities. Deriving analytic expressions for the relevant distributions we prove optimality of a conventional (s, Q)–policy for the procurement decisions. What is more, we show the cost function to have the same structure as in a traditional inventory model without returns, which allows for an easy computation of optimal control parameter values. In the subsequent sections we extend this approach to the case of general demand and return distributions. Section 7.3 provides the key result, showing that our model can be transformed into an equivalent classical inventory model without returns if procurement is controlled by an (s, S)–order policy. Optimality of an (s, S)–policy for our model is shown in Section 7.4. In Section 7.5 we use the analytic results to evaluate the impact of different return flow characteristics numerically. Section 7.6 concludes the chapter by discussing possible extensions to the model and delineating the scope of the results.

7.1 A Basic Inventory Model with Item Returns

Following the above motivation we consider a standard single item stochastic inventory model extended with a stochastic inbound item flow. For the time being, item returns are assumed to immediately raise the serviceable stock level. For an illustration one may think of situations where returned products can be reused directly without major re–processing, such as reusable packaging (compare Table 3.1). However, we remark that this assumption is for notational convenience mainly and does not limit the generality of the model essentially. In Section 7.6 we show how a recovery process involving a positive leadtime can be incorporated in this setting following a standard state–redefinition approach.

For ease of presentation we consider a discrete time setting (see Section 7.6 again for extensions to a continuous time model). We assume the following sequence of events. At the beginning of each period the inventory level is reviewed. A decision is taken on procurement orders, which are delivered after a fixed leadtime of τ periods. Subsequently, demand and returns arrive throughout the period. Any unsatisfied demand is backordered. Let

D_n^+ = demand in period n;
D_n^- = returns in period n;
D_n = $D_n^+ - D_n^-$, net demand in period n .

We assume $(D_n)_{n \in \mathbb{N}}$ to be independent identically distributed as an integer random variable D and let $p_i = \mathbb{P}\{D = i\}, i \in \mathbb{Z}$. Note that this assumption allows for stochastic dependence between demand and returns within the same period. For example, replacement demand triggered by returns can be taken into account. In contrast, there is no dependence across periods, provided that both demand and returns are themselves i.i.d. sequences. We recall from the previous chapter that the relation between the demand and return process is one of the distinguishing elements between the different models in literature. Independence or a (possibly stochastic) time lag are the two major options that have been considered. We defer a more detailed discussion concerning the appropriateness of these assumptions to Section 7.6. To describe the system's state let

Y_n = net stock at the beginning of period n before ordering;
A_n = replenishment order size in period n;
X_n = net stock at the beginning of period n after ordering and receipt of replenishments;
I_n = inventory position at the beginning of period n defined as Y_n + outstanding replenishments.

As explained above, we do not include a disposal option and therefore restrict A_n to be nonnegative for all n. Obviously, this assumption only makes sense if the average demand outweighs the average returns since otherwise the inventory level would increase to infinity.

We therefore assume

$$\mathbb{E}[D] > 0 \ . \tag{7.1}$$

The objective is to minimise the long–run average cost per time, considering fixed order costs and convex shortage and holding costs. Note that linear order costs are not relevant since any stable policy necessarily orders $\mathbb{E}[D]$ units per time on average. Let

K = fixed cost per replenishment order;
$G(x)$ = expected one period holding and backorder costs when starting with a stock level x (after ordering and receipt),

and assume $G(.)$ to be convex and $G(x) \to \infty$ for $|x| \to \infty$.

We reconsider Tables 6.2 and 6.3 to place the above model in the context of the literature reviewed in the previous chapter. As pointed out before, the above setup concurs with most of the earlier approaches in that it considers a single stock point. Hence, focus is on the behaviour of the end–item level rather than on a detailed modelling of the recovery channel. Within this perspective, the above model is rather general in that it includes both fixed costs and leadtimes, thus taking into account lotsizing as well as safety stock effects. On the other hand, it does not include a disposal option whereas a few of the previous contributions do. We discuss the tradeoff in terms of tractability between the different assumptions at the end of this chapter. The main difference with most of the single–stock models in Table 6.2 is the independence assumption of demand and returns. It is shown in this chapter that this assumption allows for fairly comprehensive results. Its justification from a practical perspective is discussed in Section 7.6. Considering the continuous–time models in Table 6.3 it should be noted that the above model encompasses batch demand and return distributions, rather than being restricted to Poisson processes. Most of all, the above setup resembles the models of Muckstadt and Isaac (1981) and Yuan and Cheung (1998). Besides the more general demand and return distributions the only difference with the former is a slightly more restrictive recovery process (see Section 7.6). The difference with the latter concerns the independence of demand and returns versus an exponentially distributed market sojourn time. Both of these contributions do not include results on optimal policy structures.

Finally, it should be noted that the above model is essentially a standard stochastic inventory model up to the difference that demand may be both positive and negative. In the subsequent sections we show that the model can be transformed into an equivalent model with nonnegative demand only.

7.2 The Unit Demand Case

We start by analysing our model for the special case of unit transaction sizes. Hence, we assume in this section that the net demand per period equals plus or minus one. This allows us to give explicit expressions for the relevant probability distributions, which simplifies the presentation of the major ideas. In the subsequent sections we follow the same approach to analyse the general demand case.

It should be noted that assuming unit transaction sizes in the above periodic review model is equivalent to a continuous review model where demand and returns are independent Poisson processes and the expected interarrival time equals the review period. Since this may be the most relevant direct application of the unit transaction case we take a continuous review perspective in this section. Hence, let us assume that demand and returns are independent Poisson processes and that a procurement order may be placed at any arrival epoch.

To comply with notational conventions we replace the period index n by a continuous argument t in this section. Moreover, let

λ_D = demand intensity;
λ_R = return intensity;
γ = return ratio defined as λ_R/λ_D;
$D(t_1, t_2)$ = net demand in time interval $[t_1, t_2)$,

and let $G(x)$ denote a cost rate per time, which is incurred while the net stock equals x. Note that assumption (7.1) implies that $\gamma < 1$. Moreover, for $\gamma = 0$ the model reduces to a conventional continuous review inventory model with Poisson demand (see, e.g., Federgruen and Zheng, 1992).

The objective is to find an order policy for this model that minimises the long–run average costs. In the conventional case, i.e. $\gamma = 0$, it is well known that it is sufficient to optimise within the class of (s, Q)–policies based on the inventory position (see, e.g. Zheng, 1991). Under this rule a replenishment order of size Q is placed whenever the inventory position drops to s. In this case the steady state distribution of $I(t)$ is uniform on $\{s+1, s+2, ..., s+Q\}$. At the end of this section we show that an (s, Q)–policy is also optimal in the case $\gamma > 0$. Therefore, we restrict our analysis to (s, Q)–policies now and in addition require s and Q to be integer. As in the conventional case we have the relation

$$X(t) = I(t - \tau) - D(t - \tau, t), \tag{7.2}$$

where I and D are independent. To characterise the long–run behaviour of X we therefore consider the steady state distributions of the inventory position and the leadtime demand.

The inventory position $I(t)$ forms a homogeneous continuous–time Markov process with state space $\mathcal{I} := \mathbb{Z} \cap [s+1, \infty)$ under the above assumptions. Note that, in contrast with traditional inventory models, the state space here is unbounded from above. The non–zero transition rates are given by

$$\lim_{\Delta_t \to 0} \frac{1}{\Delta_t} P\{I(t + \Delta_t) = k | I(t) = l\} = \begin{cases} \lambda_D & k = l - 1, \;\; l \geq s + 2 \\ \lambda_D \text{ for } k = s + Q, \, l = s + 1 \\ \lambda_R & k = l + 1, \;\; l \geq s + 1 \;\; . \end{cases}$$

Since $0 \leq \gamma < 1$ the process $I(t)$ is ergodic and is known to have the following stationary distribution (see, e.g., Muckstadt and Isaac, 1981).

Proposition 7.2.1. *Let* $0 \leq \gamma < 1$. *Then*

$$i_{s+k} := \lim_{t\to\infty} P\{I(t) = s+k\} = \begin{cases} \dfrac{1-\gamma^k}{Q} & 1 \leq k \leq Q \\ \dfrac{(\gamma^{-Q}-1)\,\gamma^k}{Q} & Q < k\ . \end{cases} \tag{7.3}$$

Proof:
It is easily verified that (7.3) satisfies the equilibrium equations of the process $I(t)$ (see also Muckstadt and Isaac, 1981) .

□

Note in particular that the steady state distribution of the inventory position is not uniform in this model. Let I_∞ denote a random variable with this probability distribution, (in the sequel, analogous notation is used for other variables). From (7.3) we get

$$\mathbb{E}[I_\infty] = s + \frac{Q+1}{2} + \frac{\lambda_R}{\lambda_D - \lambda_R} \quad \text{and} \quad \mathrm{Var}[I_\infty] = \frac{Q^2-1}{12} + \frac{\lambda_D\lambda_R}{(\lambda_D-\lambda_R)^2}\ .$$

Next let us consider the distribution of the leadtime demand.

Proposition 7.2.2. *For* $\gamma > 0$ *the distribution of the net demand during a leadtime period is given by*

$$d_k := P\{D(t-\tau,t) = k\} = e^{-(\lambda_D+\lambda_R)\tau}\sqrt{\gamma}^{(-k)}\, I_k(2\tau\sqrt{\lambda_R\lambda_D}) \quad \forall\, k \in \mathbb{Z}\ ,$$

where $I_k(2z) = z^k \sum_{l=0}^{\infty} \frac{z^{2l}}{(l+k)!k!}$ *for* $z \in \mathbb{R}$ *denotes the* k*-th modified Bessel function.*
For $\gamma = 0$ *we have* $\lim_{\lambda_R\to 0} P\{D(t-\tau,t) = k\} = e^{-\lambda_D\tau}\,(\lambda_D\,\tau)^k/\,k!$.

Proof:
We verify the expression for $k \geq 0$. For negative k the proof runs analogously. Supposing $k \geq 0$ we have

$$\begin{aligned} P\{D(t-\tau,t) = k\} &= \sum_{l=0}^{\infty} P\{D^+(t-\tau,t) = l+k\} \cdot P\{D^-(t-\tau,t) = l\} \\ &= e^{-(\lambda_D+\lambda_R)\tau}\,(\tau\lambda_D)^k \sum_{l=0}^{\infty} \tau^{2l} \frac{\lambda_D^l}{(l+k)!}\frac{\lambda_R^l}{l!} \\ &= e^{-(\lambda_D+\lambda_R)\tau}\sqrt{\gamma}^{(-k)}\, I_k(2\tau\sqrt{\lambda_R\lambda_D}). \end{aligned}$$

□

Note that $I_k(2z)$ can efficiently be approximated numerically (see Abramowicz and Stegun, 1970). Since the above distribution is independent of t we write $D(\tau)$ rather than $D(t-\tau,t)$ from now on. From the definition of D we immediately get

$$\mathbb{E}[D(\tau)] = \tau(\lambda_D - \lambda_R) \qquad \text{and} \qquad \mathrm{Var}[D(\tau)] = \tau(\lambda_D + \lambda_R) \ .$$

By (7.2) the above results can be used to characterise the steady state distribution of $X(t)$, though not in an easy, closed form expression. For the first two moments we get

$$\mathbb{E}[X_\infty] = s + \frac{Q+1}{2} + \frac{\lambda_R}{\lambda_D - \lambda_R} - \tau(\lambda_D - \lambda_R) \tag{7.4}$$

$$\mathrm{Var}[X_\infty] = \frac{Q^2-1}{12} + \frac{\lambda_D \lambda_R}{(\lambda_D - \lambda_R)^2} + \tau(\lambda_D + \lambda_R) \ . \tag{7.5}$$

Note that for fixed control parameters both the mean and the variance of the net inventory in steady state increase with λ_R.

We now have all the prerequisites to analyse the behaviour of the cost function. Let $C(s,Q)$ denote the long–run expected average costs per time for given control parameters s and Q. In the usual way, by considering individual replenishment cycles and applying the renewal reward theorem (see, e.g., Hadley and Whitin, 1963) we have

$$C(s,Q) = K\,\frac{\lambda_D - \lambda_R}{Q} + \mathbb{E}[G(X_\infty)] \ . \tag{7.6}$$

By virtue of relation (7.2) we can rewrite this expression in terms of the inventory position. To this end, let $\tilde{G}(l)$ denote the expected inventory cost rate at time $t+\tau$ when the inventory position at time t equals l, i.e.

$$\tilde{G}(l) := \mathbb{E}[G(X(t+\tau))\,|\,I(t) = l] = \mathbb{E}[G(l - D(\tau))].$$

Using (7.3), the cost function can then be written as

$$\begin{aligned} C(s,Q) &= K\,\frac{\lambda_D - \lambda_R}{Q} + \mathbb{E}[\,G(I_\infty - D(\tau))] \\ &= K\frac{\lambda_D - \lambda_R}{Q} + \sum_{l=s+1}^{\infty} i_l\,\tilde{G}(l) \\ &= \frac{1}{Q}[c + \sum_{l=1}^{Q}(1-\gamma^l)\tilde{G}(s+l) + (\gamma^{-Q}-1)\sum_{l=Q+1}^{\infty}\gamma^l\,\tilde{G}(s+l)] \ , \end{aligned} \tag{7.7}$$

where $c := K(\lambda_D - \lambda_R)$. However, in contrast with the conventional model, this expression involves an infinite sum with coefficients depending on the control parameters s and Q. This complicates the analysis of $C(s,Q)$ in this form for the return flow model. The following proposition shows how this difficulty can be overcome by rewriting the cost function further.

Proposition 7.2.3. *Let*

$$H(k) := (1-\gamma)\sum_{l=0}^{\infty}\gamma^l \tilde{G}(k+l)\ . \tag{7.8}$$

Then $H(k)$ is convex in k and

$$C(s,Q) = [c + \sum_{k=s+1}^{s+Q} H(k)]\ /\ Q\ . \tag{7.9}$$

Proof:
Inserting (7.8) into the righthandside of (7.9) yields

$$[c + \sum_{k=s+1}^{s+Q} H(k)]\ /\ Q$$

$$= [c + (1-\gamma)\sum_{k=s+1}^{s+Q}\sum_{l=0}^{\infty}\gamma^l \tilde{G}(k+l)\ ]\ /\ Q \quad =$$

$$= \{c + (1-\gamma)\ [\sum_{l=1}^{Q}\tilde{G}(s+l)\sum_{k=1}^{l}\gamma^{k-1} + \sum_{l=Q+1}^{\infty}\tilde{G}(s+l)\sum_{k=1}^{Q}\gamma^{l-k}\ ]\ \}\ /\ Q$$

$$= [c + \sum_{l=1}^{Q}(1-\gamma^l)\tilde{G}(s+l) + (\gamma^{-Q}-1)\sum_{l=Q+1}^{\infty}\gamma^l\,\tilde{G}(s+l)\ ]\ /\ Q\ ,$$

which is equal to $C(s,Q)$ according to (7.7). Moreover, $\tilde{G}(l) = \mathbb{E}[G(l-D)]$ is convex due to the convexity of G and therefore $H(k)$ is convex as a sum of convex functions. □

The advantage of expression (7.9) is that it has exactly the same structure as the cost function in a conventional (s,Q)–model. Therefore, we can apply the sequential procedure proposed by Federgruen and Zheng (1992) to compute the optimal control parameters. This result is summarised in the following corollary.

Corollary 7.2.1. *The optimal control parameters s and Q of the above inventory model with return flow can be found in a linear search by successively evaluating the minimum costs $C^*(Q) := C(s^*(Q), Q)$ for increasing Q, where $s^*(Q)$ denotes the optimal reorder level for given order quantity Q. The iteration starts with $s^*(1) = \arg\min_{k\in\mathbb{Z}} H(k)$ and exploits the fact that $s^*(Q) - 1 \leq s^*(Q+1) \leq s^*(Q)$. Moreover, $-C^*(Q)$ is unimodal and the search can be stopped as soon as $C^*(Q+1) \geq C^*(Q)$.*

Proof:
The proof follows directly from Proposition 7.3 and the results from Federgruen and Zheng (1992). □

Equations (7.8) and (7.9) show how a stochastic item return flow can be incorporated into the setting of a traditional inventory model. Replacing the conditioned inventory cost rate function $\tilde{G}(.)$ by a moving–average like modification turns out to be key. This relationship may be further interpreted as follows. Suppose that returned and new items are kept in two distinct inventories (with the same cost parameters). Demand is assumed to be served with preference from returned items. Only if there are no returned items on–hand, demand is served from new items. Moreover, suppose that procurement orders are controlled according to an (s, Q)–policy based on an inventory position of new items only. We refer to this model as 'two–bucket model' while the model considered so far is simply called 'return flow model'. The following proposition shows that both models are equivalent and that $H(.)$ is the inventory cost rate in the 'two–bucket' model conditioned on the new inventory position.

Proposition 7.2.4. *Let $\tilde{C}(s, Q)$ denote the long–run expected average costs for given control parameters s and Q in the above 'two bucket' inventory model. Then*

$$\tilde{C}(s, Q) = C(s, Q) = [c + \sum_{k=s+1}^{s+Q} H(k)] \; / \; Q$$

and $H(.)$ is the expected inventory cost rate conditioned on the inventory position of new items.

Proof:
Define inventory positions $K(t)$ and $L(t)$ for new and returned items respectively via the following updating rules:

each return triggers a transition $(K, L) \to \quad (K, L+1)$

each demand triggers a transition $(K, L) \to \begin{cases} (K, L-1) & L > 0 \\ (K-1,\, 0) \text{ if } & L = 0, K > s+1 \\ (s+Q,\, 0) & L = 0, K = s+1 \end{cases}$

See Figure 7.1 for a graphical representation.

Clearly, we have $K(t) + L(t) = I(t) + \kappa$, with some constant κ. Without loss of generality we may choose κ equal to zero. Moreover, since the cost parameters are identical the two models are indeed equivalent and we have $\tilde{C}(s, Q) = C(s, Q)$.

To interpret the function $H(.)$ in the setting of the two–bucket model, note that $(K, L)(t)$ is a two-dimensional Markov process on $\mathbb{Z} \cap [s+1, s+Q] \times \mathbb{Z}_{\geq 0}$ in continuous time with non–zero transition rates

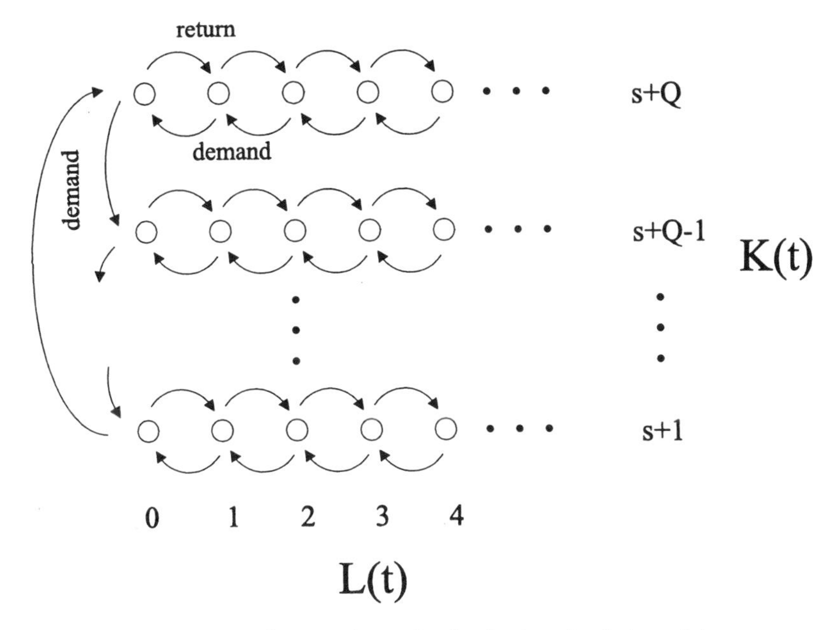

Fig. 7.1. System dynamics in the two-bucket model

$$\lim_{\Delta_t \to 0} \frac{1}{\Delta_t} P\{(K,L)(t+\Delta_t) = (k,l) | (K,L)(t) = (m,n)\}$$
$$= \begin{cases} \lambda_D \text{ for } k = m-1,\ m \geq s+2,\ l = n = 0 \\ \lambda_D \quad k = s+Q,\ m = s+1,\ l = n = 0 \\ \lambda_D \quad k = m, \quad l = n-1,\ n \geq 1 \\ \lambda_R \quad k = m, \quad l = n+1,\ n \geq 0\ . \end{cases} \tag{7.10}$$

We show that the steady state distributions of $K(t)$ and $L(t)$ are independent. Solving the equilibrium equations for (7.10) gives

$$\pi_{kl} := \lim_{t\to\infty} P\{(K,L)(t) = (k,l)\} \;=\frac{1-\gamma}{Q}\,\gamma^l \quad \text{for } s+1 \leq k \leq s+Q,\ 0 \leq l\,,$$

which implies

$$\rho_k := \lim_{t\to\infty} P\{K(t) = k\} = \frac{1}{Q} \quad \text{for } s+1 \leq k \leq s+Q$$

and

$$q_l := \lim_{t\to\infty} P\{L(t) = l\} = (1-\gamma)\,\gamma^l \quad \text{for } l \geq 0\,.$$

Thus, $\pi_{kl} = \rho_k\, q_l$ and the two processes are independent. Moreover, note that $(\rho_k)_k$ is independent of γ and $(q_l)_l$ independent of s and Q. Now we can write

$$\begin{aligned}\tilde{C}(s,Q) &= C(s,Q) \\ &= \frac{c}{Q} + \frac{1}{Q}\sum_{k=s}^{s+Q} H(k) \\ &= \frac{c}{Q} + \sum_{k=s}^{s+Q} \rho_k H(k) \ .\end{aligned}$$

Hence, H(.) is the conditioned inventory cost rate in the two–bucket model. □

Proposition 7.5 shows that the inventory position in the return flow model can be decomposed into two independent parts, one being independent of the return flow and the other independent of the control parameters. The return flow only affects the average inventory – and hence the average cost rate – between two transitions of the new item inventory position. In Section 7.3 we show that this result also holds in a more general context. We conclude the analysis of the unit demand case by showing optimality of (s,Q)–order policies.

Proposition 7.2.5. *There exists an (s,Q)–order policy for the above inventory model with return flow that is average cost optimal among all history-dependent order policies with decision epochs at the moments of demand or return occurrence.*

Proof:
The proof runs analogous with the traditional case, i.e $\gamma = 0$. We only sketch the main steps. A more detailed proof for the general demand case is given in Section 7.4. Due to relation (7.2) the total cost function only depends on the inventory position and not on the quantities ordered at individual epochs, just as in the conventional model (see, e.g., Arrow et al., 1958). Hence, it suffices to consider control policies based on the inventory position. Moreover, as discussed above, the model is equivalent to a discrete time model by considering the system state at transition epochs. For the discrete model general results on average cost Markov decision problems (see, e.g., Sennott, 1989) assure the existence of a stationary average cost optimal policy. Average cost optimality of an (s,Q)–policy follows since any stationary policy based on the inventory position in the above model is equal to an (s,Q)–policy up to a transient phase. □

7.3 General Demand Case: Analysis of the Cost Function

Let us now return to the general demand, periodic review case as introduced in Section 7.1. Two main difficulties complicate the development, compared to the unit demand case. On the one hand, the analysis of the cost function is more involved since the relevant probability distributions can, in general, not be calculated explicitly. On the other hand, more effort is required for establishing the optimal policy structure since a stationary policy is not automatically a critical number policy in this case.

For notational convenience we assume that τ equals zero, i.e. replenishment orders are delivered immediately. A positive leadtime can easily be incorporated in the same way as in the previous section. We defer a more detailed discussion to Section 7.6 together with the introduction of an explicit recovery process. For the case without item returns, i.e. positive demand only, it is well known that the long–run average costs are minimized by an (s, S)–order policy (see, e.g., Zheng, 1991). Under this rule a replenishment order is placed so as to bring the inventory level back to a target level S whenever it drops to or below the reorder level s. In the next section we prove this class of policies to be optimal also for our return flow model. In this section we first analyse the structure of the cost function for a given (s, S)–policy. To this end, we make use of a similar decomposition approach as for the unit demand case. The main result of this section is a transformation of the return–flow model into an equivalent standard inventory model without returns.

Hence, assume that replenishment orders are controlled according to an (s, S) policy. In this case the begin of period stock level after possible ordering evolves as follows.

$$x_{n+1} = \begin{cases} x_n - d_n & \text{if } d_n < x_n - s \\ S & \text{else} \end{cases}$$

Recall that d_n may be negative. Therefore, $(X_n)_{n \in I\!N} \subset [s+1, \infty)$ is again unbounded from above, in contrast with traditional inventory models. Similar to Figure 7.1 we can split the process (X_n) in two parts such that the portion capturing the return flow is independent of the control parameters. To this end let $\mathcal{S} := \{s+1, \ldots, S\}$ and define processes $(K_n), (L_n)$ recursively by

$$k_0 := x_0,\; k_{n+1} := \begin{cases} k_n - \max\{0, d_n - l_n\} & \text{if } d_n < k_n + l_n - s \\ S & \text{else,} \end{cases}$$

$$l_0 := 0, \;\; l_{n+1} := \max\{0, l_n - d_n\} \; .$$

Figure 7.2 shows a realisation of these processes. Note that the dynamics are the same as in the 'two–bucket' model introduced in Proposition 7.5. Therefore, one may again interpret L_n as stock that is built up by returns whereas K_n denotes newly procured items. Again, demand is served with returned items first.

Considering Figure 7.2 it turns out that all particularities induced by the return flow are captured by the process (L_n) whereas (K_n) behaves like a standard inventory process. Moreover, it is intuitively clear from the figure that the processes (K_n) and (L_n) are independent in the long–run. This

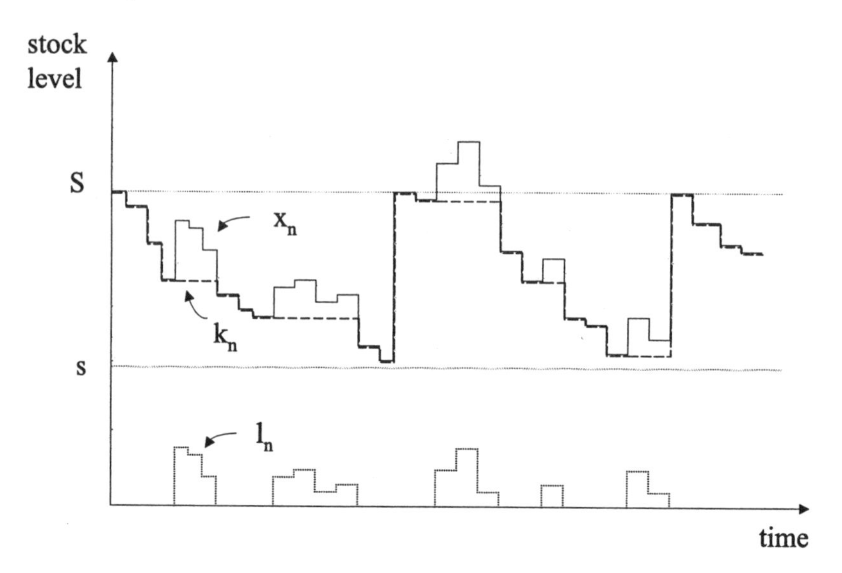

Fig. 7.2. Decomposition of the inventory process

allows us to reduce the analysis of the inventory process (X_n) to an analysis of K_n and to incorporate L_n in the cost coefficients. The approach is similar to a busy period analysis in priority queueing models.

To make these observations rigorous, note that $x_n = k_n + l_n$ for all n and that (K_n, L_n) is a two-dimensional Markov process on $\mathcal{I} := \mathcal{S} \times \mathbb{N}$ with transition matrix $\mathbb{P}_{l,l'}^{k,k'} := \mathbb{P}\{K_{n+1} = k', L_{n+1} = l' | K_n = k, L_n = l\}$ given by

$$\mathbb{P}_{l,l'}^{k,k'} = \mathbb{1}_{k=k'} p_{l-l'} + \mathbb{1}_{l'=0} [\mathbb{1}_{k>k'} p_{l+k-k'} + \mathbb{1}_{k'=S} \sum_{i \geq 0} p_{l+k-s+i}] \ ,$$

for $k, k' \in \mathcal{S}$, $l, l' \in \mathbb{N}$ and $\mathbb{1}_{A(x)}$ denoting the indicator function of the set $A(x)$. The next two Lemmas show that X_n has a stationary distribution in product form.

To this end, first note that (L_n) is a random walk with drift on the positive half line. The corresponding transition matrix is

$$\mathbb{Q}_{l,l'} := \mathbb{P}\{L_{n+1} = l' | L_n = l\} = \mathbb{1}_{l' \neq 0} p_{l-l'} + \mathbb{1}_{l'=0} \sum_{i \geq 0} p_{l+i}$$

Due to assumption (7.1) this process converges:

Lemma 7.3.1. *The random walk (L_n) is ergodic. In particular (L_n) admits a unique stationary distribution $(q_l)_{l \geq 0}$ that is independent of s and S.*

Proof:
Ergodicity of the random walk under the condition $\mathbb{E}[D] > 0$ has been shown

by Meyn and Tweedie (1993, p.270). Since the process (L_n) does not depend on the control parameters s and S the same holds for its limiting distribution.

□

Next, we show independence of the limit behaviour of (L_n) and (K_n).

Lemma 7.3.2. *Let $\pi(k,l)$ be a $\mathbb{P}$–invariant probability distribution on $\mathcal{I}$. Then for any fixed $k \in \mathcal{S}$, $\pi(k,.)$ is a $\mathbb{Q}$–invariant measure on $\mathbb{N}$.*

Proof:
First assume $l' > 0$. Then we have

$$\begin{aligned}
\pi(k',l') &= \sum_{k\in\mathcal{S}} \sum_{l\geq 0} \pi(k,l) \mathbb{P}_{l,l'}^{k,k'} \\
&= \sum_{k\in\mathcal{S}} \sum_{l\geq 0} \pi(k,l) \mathbb{1}_{k=k'} p_{l-l'} \\
&= \sum_{l\geq 0} \pi(k',l) p_{l-l'} \\
&= \sum_{l\geq 0} \pi(k',l)\, \mathbb{Q}_{l,l'}.
\end{aligned}$$

To complete the proof for the case $l = 0$ let $\rho_k = \sum_{l\geq 0} \pi(k,l)$ for $k \in \mathcal{S}$. Then

$$\begin{aligned}
\pi(k',0) &= \rho_{k'} - \sum_{l'>0} \pi(k',l') \\
&= \rho_{k'} - \sum_{l'>0} \sum_{l\geq 0} \pi(k',l)\, \mathbb{Q}_{l,l'} \\
&= \rho_{k'} - \sum_{l'\geq 0} \sum_{l\geq 0} \pi(k',l)\, \mathbb{Q}_{l,l'} + \sum_{l\geq 0} \pi(k',l)\, \mathbb{Q}_{l,0} \\
&= \rho_{k'} - \sum_{l\geq 0} \pi(k',l) + \sum_{l\geq 0} \pi(k',l)\, \mathbb{Q}_{l,0} \\
&= \sum_{l\geq 0} \pi(k',l)\, \mathbb{Q}_{l,0}
\end{aligned}$$

□

Corollary 7.3.1. *If $\pi(k,l)$ is a $\mathbb{P}$–invariant probability distribution on $\mathcal{I}$ then there exists a probability distribution (ρ_k) on $\mathcal{S}$ such that $\pi(k,l) = \rho_k q_l$ for all $k \in \mathcal{S}$ and all $l \in \mathbb{N}$.*

Proof:
For $k \in \mathcal{S}$ define $\rho_k = \sum_{l\geq 0} \pi(k,l)$. For $\rho_k = 0$ the proof is evident. Hence,

assume $\rho_k > 0$. Due to Lemma 7.8 $\pi(k,l)/\rho_k$ is a $\mathbb{Q}$–invariant probability distribution on $\mathbb{N}$. According to Lemma 7.7 the invariant distribution of $\mathbb{Q}$ is unique. Therefore, $\pi(k,l)/\rho_k = q_l$ for all $l \in \mathbb{N}$ which completes the proof. □

Corollary 7.9 provides us with the prerequisites to transform the return–flow model into an equivalent inventory model without return flow. To this end, consider a standard (s,S)–inventory model with demand distribution $(\tilde{p}_i)_{i\in\mathbb{N}}$ defined by

$$\tilde{p}_0 := \sum_{l\geq 0} q_l \sum_{j\geq 0} p_{l-j} \quad \text{and} \quad \tilde{p}_i := \sum_{l\geq 0} q_l p_{i+l} \quad \text{for } i \geq 1, \tag{7.11}$$

fixed order costs K and expected one–period holding and backorder costs

$$H(k) := \sum_{l\geq 0} q_l G(k+l) \tag{7.12}$$

for a given inventory level $k \in \mathcal{S}$. Note that $H(k)$ may be infinite. However, we show in the next section that infinity of $H(k)$ for some k implies infinite average costs for any control policy. Denote the long–run average costs per time for a given (s,S)-policy in this model by $\tilde{C}(s,S)$. It is well known (see, e.g., Federgruen and Zheng, 1992) that in this standard model the limiting probability distribution of the stock–level is given by the unique solution $(\tilde{\rho}_k)_{k\in\mathcal{S}}$ of the equation

$$\rho_{k'} = \sum_{k\geq k'} \rho_k \tilde{p}_{k-k'} + 1\!\!1_{k'=S} \sum_{k} \rho_k \sum_{i\geq 0} \tilde{p}_{k-s+i} \quad \text{for } k' \in \mathcal{S} . \tag{7.13}$$

The next Lemma shows that $(\tilde{\rho}_k)$ is the limiting distribution of the process (K_n).

Lemma 7.3.3. *The Markov chain (K_n, L_n) is ergodic. In particular, $\pi(k,l) := \tilde{\rho}_k q_l$ is the unique invariant distribution of $\mathbb{P}$ and has a product form.*

Proof:
Let $\pi(k,l)$ be an arbitrary probability distribution on $\mathcal{I}$. Due to Corollary 7.9 it suffices to consider distributions of the form $\pi(k,l) = \rho_k q_l$, with a probability distribution (ρ_k) on $\mathcal{S}$ and (q_l) from Lemma 7.7. π is $\mathbb{P}$–invariant if and only if

$$\rho_{k'} q_{l'} = \sum_{k\in\mathcal{S}} \sum_{l\geq 0} \rho_k q_l \mathbb{P}^{k,k'}_{l,l'} \quad \forall\ k' \in \mathcal{S}\ ,\ l' \in \mathbb{N}\ .$$

For $l' > 0$ this equation holds independent of (ρ_k) as seen in the proof of Lemma 7.8. For the case $l' = 0$ we have the following equivalences:

$$\begin{aligned}
\rho_{k'} q_0 &= \sum_{k \in \mathcal{S}} \sum_{l \geq 0} \rho_k q_l \mathbb{P}^{k,k'}_{l,0} \\
\iff \quad \rho_{k'} &= \sum_{k \in \mathcal{S}} \rho_k \sum_{l \geq 0} q_l \mathbb{P}^{k,k'}_{l,0} + \rho_{k'}(1 - q_0) \\
&= \sum_{k \in \mathcal{S}} \rho_k 1\!\!1_{k \geq k'} \sum_{l \geq 0} q_l p_{l+k-k'} + \sum_{k \in \mathcal{S}} \rho_k 1\!\!1_{S=k'} \sum_{i \geq 0} \sum_{l \geq 0} q_l p_{l+k-s+i} \\
&\quad + \rho_{k'} \sum_{l \geq 0} q_l \sum_{j \geq 1} p_{l-j} \qquad (7.14) \\
&= \sum_{k \geq k'} \rho_k \tilde{p}_{k-k'} + 1\!\!1_{k'=S} \sum_{k \in \mathcal{S}} \rho_k \sum_{i \geq 0} \tilde{p}_{k-s+i} \\
\iff \quad \rho_{k'} &= \tilde{\rho}_{k'} \quad \forall k' \in \mathcal{S}\,, \qquad (7.15)
\end{aligned}$$

where (7.14) uses $\mathbb{Q}$–invariance of (q_l) and (7.15) follows from (7.13). This completes the proof. □

This leads us to the main result of this section:

Proposition 7.3.1. *For any given (s, S)–policy the return–flow model and the above standard inventory model yield the same long–run average costs per time, i.e.*

$$C(s, S) = \tilde{C}(s, S) \quad \forall s < S$$

Proof:
We use the split–variable formulation $(X_n) = (K_n + L_n)$ for the stock–level in the return–flow model. From the stationary distributions of (K_n) and (L_n) we get

$$\begin{aligned}
C(s, S) &= \sum_{k \in \mathcal{S}} \sum_{l \geq 0} \rho_k q_l \left[G(k + l) + K \sum_{i \geq 0} p_{k-s+l+i}\right] \\
&= \sum_{k \in \mathcal{S}} \rho_k [H(k) + K \sum_{i \geq 0} \tilde{p}_{k-s+i}] = \tilde{C}(s, S).
\end{aligned}$$

□

As in the previous section, the above results make the machinery of classical inventory theory available for the return flow model. In particular, standard algorithms may be used for computing optimal control parameters. Applications are discussed in more detail in Section 7.5.

7.4 General Demand Case: Optimal Policy Structure

In this section we show that an (s, S)–order policy is average cost optimal in the return flow model. For traditional inventory models optimality of an

(s, S)–policy under the condition of backordering unsatisfied demand is well known. A remarkably direct proof has been given by Zheng (1991). His approach relies on considering a relaxed model including disposal, for which optimality of an (s, S)–order policy follows from deriving a bounded solution of the optimality equation of the corresponding Markov decision process. Optimality in the original model follows since both models differ in at most one period. In the return flow case this last conjecture fails since the state space is unbounded from above for any order policy. Therefore, we follow a more classical approach here, based on the limit behaviour of a discounted cost model. To this end, we exploit general theory of Markov decision processes that has been well developed in the past two decades. In particular, we make use of Sennott's results on infinite state Markov decision processes with unbounded costs (Sennott, 1989).

We proceed in three steps as follows. First, we note that optimality of a nonstationary (s, S)–policy for the finite horizon discounted cost case follows from Scarf's well–known K-convexity proof. Then we show convergence of the finite horizon to the infinite horizon discounted costs. Finally, analysing the limit behaviour for discount factors approaching one yields an optimal (s, S)–policy for the average cost model.

In the sequel we consider our model in terms of Markov decision processes with state variable Y_n and decision variable A_n as defined in Section 7.1. Note that the quantity ordered in any state $y \in \mathbb{Z}$ may be bounded by $\bar{y} - y$ for any $\bar{y}$ with $G(\bar{y}) > G(y^*) + K$, where y^* is a minimiser of G. Any policy ordering more than this quantity is dominated by first ordering upto y^* (if $y < y^*$, otherwise do not order at all) and postponing the remainder of the order. Hence, we may assume finite action sets without loss of generality. In the sequel we use the following standard notation. Let

$V_\alpha^N(y)$ = infimum of the expected N–period α–discounted costs given an initial stock level y;

$V_\alpha(y)$ = infimum of the expected infinite horizon α–discounted costs given an initial stock level y.

In his seminal paper Scarf (1960) proves optimality of a nonstationary (s, S)–order policy for a discounted cost finite horizon inventory model with backordering and K–convex one–period expected costs. He shows by induction that K–convexity of the expected costs over n periods implies optimality of an (s, S)–rule in period $N - n$ which again yields K–convexity of the $n+1$–period expected costs. It has been pointed out earlier that Scarf's proof does not require positive demand (Heyman and Sobel, 1984). In fact, the entire proof remains valid for i.i.d. demand realisations admitting both positive and negative values. Since K–convexity of the one–period cost function holds in our return–flow model (as we have assumed $G(.)$ to be even convex) we get optimality of an (s, S)–policy for the finite horizon total discounted cost criterion.

In another seminal contribution to inventory control theory Iglehart (1963) extends Scarf's results to an infinite horizon setting. We follow this approach for the return flow model. By exploiting more recent advances in the theory of Markov decision processes a slightly more compact proof is given. For this step we require an additional assumption. As pointed out in the previous section the function $H(k)$ defined in (7.12) is not necessarily finite. Note that by Proposition 7.11 and standard results in traditional inventory models (see, e.g., Zheng, 1991) $H(k)$ is the expected average cost per time incurred during a first passage from inventory level k to or below level $k-1$ when no orders are placed. Therefore, if $H(k)$ is infinite for all k, any order policy yields infinite long–run average costs in the return flow model. Since in this case the problem is somewhat ill–posed it makes sense to require $H(k)$ to be finite for at least one k. Basic algebraic manipulation shows that $H(k)$ must then be finite for all k as follows. Using $\mathcal{Q}$–invariance of (q_l) we have

$$
\begin{aligned}
H(k) &= q_0 G(k) + \sum_{l=1}^{\infty}\sum_{l'=0}^{\infty} q_{l'} p_{l'-l} G(k+l) \\
&= q_0 G(k) + \sum_{i=-\infty}^{\infty} p_{k-i} \sum_{l'=\max(0,k-i+1)}^{\infty} q_{l'} G(l'+i) \,.
\end{aligned}
$$

Note that the incomplete sum

$$
\sum_{l'=\max(0,k-i+1)}^{\infty} q_{l'} G(l'+i) = H(i) - \sum_{l'=0}^{k-i} q_{l'} G(l'+i)
$$

is infinite if $H(i)$ is so. Hence, $H(i)=\infty$ implies $H(i+l)=\infty$ for all l with $p_l > 0$. Convexity of $G(.)$ then yields $H(k)=\infty$ for all k. This motivates the following assumption:

$$
H(k) < \infty \quad \forall\ k \in \mathbb{Z}. \tag{7.16}
$$

We can now use Sennott's results to show convergence of the finite horizon discounted cost setting:

Lemma 7.4.1. *For every $0 < \alpha < 1$ there exists an (s,S)–order policy that is α-discounted cost optimal for all starting states y in the infinite horizon return flow model.*

Proof:
Sennott (1989) shows convergence of successive approximation to the infinite horizon discounted cost function under the condition that $V_\alpha(y)$ is finite for all y and α. Moreover, this condition is shown to hold if there exists a stationary policy inducing an ergodic Markov chain and yielding finite average costs in steady state. In the return flow model any (s,S)–policy induces an ergodic Markov chain as shown in Section 7.3. Moreover, by Proposition 7.11 we have

$C(S-1, S) \leq H(S) + K$ which is finite for all S due to (7.16). Hence, we have

$$\lim_{N \to \infty} V_\alpha^N(y) = V_\alpha(y) \text{ for } y \in \mathbb{Z}, \ 0 < \alpha < 1.$$

Therefore, using Scarf's argument, $V_\alpha(.)$ is again K–convex and satisfies the optimality equation

$$V_\alpha(y) = \min_{a \geq 0}\{1\!\!1_{a>0} K + G(y+a) + \alpha \sum_{i=-\infty}^{\infty} p_j V_\alpha(y+a-i)\}, \qquad (7.17)$$

which implies discounted cost optimality of an (s, S)–order policy. □

It remains to be shown that the solution of the discounted cost model converges to an average cost optimal policy for $\alpha \to 1$. Sufficient conditions have been given by Sennott (1989) for general Markov decision processes. We show that these conditions hold in our return flow model.

Lemma 7.4.2. *Let $(\alpha_n)_{n \in \mathbb{N}}$ be a sequence of discount factors converging to 1 and let $(s_{\alpha_n}, S_{\alpha_n})$ be a sequence of corresponding discounted cost optimal policies in the return flow model. There exists an accumulation point (s_1, S_1) of the policy sequence and any accumulation point yields an average cost optimal policy.*

Proof:
Sennott shows this convergence result for general infinite state Markov decision processes satisfying three conditions:

(i) $V_\alpha(y) < \infty$ for all y and α;
(ii) there exist $y_0 \in \mathbb{Z}$ and a nonnegative N such that $h_\alpha(y) := V_\alpha(y) - V_\alpha(y_0) \geq -N$ for all y and α;
(iii) for every y there exists a nonnegative M_y such that $h_\alpha(y)$ from (ii) satisfies $h_\alpha(y) \leq M_y$ and there exists an action $a(y)$ such that $\sum_z P_{yz}(a(y))\, M_z < \infty$.

We verify these conditions in the return flow model. (i) has already been shown in the proof of Lemma 7.12.

For (ii), note that $h_\alpha(y) \geq -K$ if $y_0 < S_\alpha$ as a consequence of (7.17). Hence, it suffices to show that (S_{α_n}) is bounded from below. Let $S_H = \arg\min H(y)$. Since $G(y) \to \infty$ for $|y| \to \infty$ there exists an $\underline{S}$ such that $G(y) > H(S_H)/(1-\tilde{p}_0)$ for all $y \leq \underline{S}$. We show that $\underline{S} < S_{\alpha_n}$ for all n.

Assume that $s_\alpha < S_\alpha \leq \underline{S}$ for some $\alpha = \alpha_n$. Let $\Delta = S_H - S_\alpha$ and let $c_\alpha(s_\alpha, S_\alpha; y)$ be the expected alpha–discounted costs in the return flow model when using policy (s_α, S_α) and starting in level y. By considering cycle costs we show that $c_\alpha(s_\alpha, S_\alpha; s_\alpha) > c_\alpha(s_\alpha + \Delta, S_\alpha + \Delta; s_\alpha)$ in contradiction to optimality of (s_α, S_α).

Analogous with $H(y)$ let $H_\alpha(y)$ denote the expected alpha–discounted costs incurred during a first passage from inventory level y to or below level

$y-1$ when no orders are placed. Note that for each sample path starting in s_α the inventory levels under policies (s_α, S_α) and $(s_\alpha + \Delta, S_\alpha + \Delta)$ differ by exactly Δ in all periods (except for the starting state). It is therefore sufficient to show that $H_\alpha(y) > H_\alpha(y+\Delta)$ for all $y \le S_\alpha$.
We have

$$H_\alpha(y+\Delta) \le \frac{1}{1-\tilde{p}_0} H(y+\Delta) = \frac{1}{1-\tilde{p}_0} \sum_{l \ge 0} q_l G(y+\Delta+l) \ \forall\, y \quad (7.18)$$

$$< G(y) \quad \text{for } y \le S_\alpha \le \underline{S} \quad (7.19)$$

$$\le H_\alpha(y) \quad \forall\, y\,, \quad (7.20)$$

where (7.18) follows from the monotonicity of costs in α, (7.19) uses the definition of $\underline{S}$ and Δ and convexity of $G(.)$, and (7.20) holds since cycle costs dominate the one period costs for all α. This completes the proof of (ii).

Choosing $y_0 < \underline{S}$ in accordance with (ii) yields $h_\alpha(y) \le K$ for $y < y_0$. Hence, in (iii) we may chose $M_y = K$ for all $y < y_0$ and it suffices to show (iii) with sums restricted to $z > y_0$. Sennott shows that (iii) holds if there exists a stationary policy inducing an irreducible, ergodic Markov chain with finite average costs in steady state. In this case M_y can be chosen as the expected cost of a first passage from y to y_0 under the given policy. As we have seen in the previous section any (s, S)–policy induces an ergodic Markov chain on $[s+1, \infty) \cap \mathbb{Z}$ in the return flow model and yields finite average costs according to (7.16). Consider the policy $(y_0 - 1, y_0)$ and choose M_y as above. If $q_l > 0$ for all $l \ge 0$ the induced chain is also irreducible and we are done. If $q_l = 0$ for some l, Sennott's proof shows $\sum_z P_{yz}(a(y))M_z < \infty$ for all recurrent states y. Convexity of $G(.)$ then assures this sum to be finite also for the transient states. This shows condition (iii) and completes the proof. □

This completes the optimality proof of (s, S)–order policies in the return flow model. We summarise the results in the following proposition.

Proposition 7.4.1. *There exists a stationary (s, S)–order policy that is average cost optimal among all order policies in the return flow model.*

Proof:
The proof is an immediate consequence of Lemmas 7.12 and 7.13. □

7.5 Numerical Examples

The results in the previous sections show that the proposed return flow inventory model bears essentially the same mathematical structure as a conventional (s, S)–model. From a managerial perspective another important

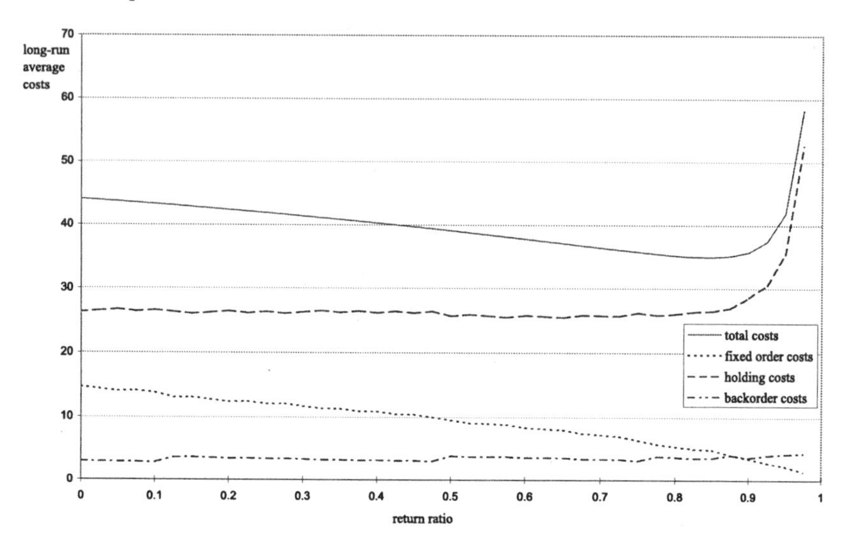

Fig. 7.3. Cost impact of the return ratio

question is the quantitative impact of the goods returns on the expected system costs. To address this issue we have carried out a number of numerical experiments exploiting the above results. This section summarises the major results grouped around three issues, namely (i) the variation of the system costs in terms of the return flow, (ii) the performance of simple heuristic policies, and (iii) the value of enhanced return information.

Throughout this section we make use of the following example of a periodic review system. Demand and returns per period are Poisson distributed with mean λ_D and λ_R respectively. (Note that this choice allows to characterise the return distribution by a single parameter.) Replenishment orders are delivered after a fixed leadtime τ. The cost function encompasses fixed order costs K per order and linear holding and backorder costs c_h and c_b per item per period. In what follows we take $\lambda_D = 10$, $\tau = 5, K = 50, c_h = 1, c_b = 100$, while λ_R is varied in the interval $[0, \lambda_D)$.

Some general remarks may be in order, concerning the numerical implementation of the results of the previous sections. For the computation of both the transformed cost function $H(.)$ and the transformed demand probability distribution $(\tilde{p}_i)_{i\geq 0}$ in (7.11) the stationary distribution (q_i) of the random walk (L_n) is required. In general, (q_i) will be computed by numerically solving the stationarity equation for $\mathcal{Q}_{l,l'}$. However, due to the infinite state space some additional considerations are required. In the case of finite demand batch sizes, i.e. $p_i = p_{-i} = 0$ for all $i > i_{\max}$ for some integer $i_{\max}$, an efficient approximation scheme can be used exploiting geometric tail behaviour of the stationary distribution (see, e.g., Tijms, 1994). In this way matrix dimensions assuring good convergence accuracy can be kept small. Using generating functions it is shown that $lim_{l\to\infty} q_l/q_{l+1} = \eta$ where the

Table 7.1. Optimal replenishment policy for varying return ratio

Return ratio	Optimal policy s^*	S^*	S^*-s^*	Long–run average costs total	fixed	holding	backorder
0.0	59	88	29	44.1	14.7	26.3	3.0
0.1	55	83	28	43.3	13.8	26.7	2.8
0.2	50	78	28	42.4	12.4	26.5	3.5
0.3	46	72	26	41.4	11.7	26.3	3.3
0.4	42	66	24	40.3	10.9	26.2	3.2
0.5	37	60	23	39.1	9.5	25.7	3.9
0.6	33	54	21	37.8	8.3	25.9	3.6
0.7	29	47	18	36.4	7.2	25.8	3.4
0.8	24	40	16	35.2	5.4	26.1	3.7
0.9	19	31	12	35.8	3.5	28.7	3.6
0.975	12	19	7	58.2	1.4	52.6	4.3

constant η is obtained as the smallest root outside the unit circle of the polynomial $r(x) = x^N - \sum_{i=0}^{2N} p_{N-i}x^i$. For infinite batch sizes some additional truncation or rounding is required.

Let us now consider the impact of the goods return flow on the expected system costs. To this end, we compute a pair of optimal control parameters (s^*, S^*) and the corresponding average costs in the above example for a sequence of values of λ_R. As explained in Section 7.3, we can use standard optimisation algorithms for this purpose. Specifically, we apply the well–known method due to Zheng and Federgruen (1991), which relies on a linear search in the (s, S)–plane limited by upper and lower bounds.

Figure 7.3 shows the optimised system costs $C(s^*, S^*)$ and a decomposition into the different cost components as a function of the return ratio $\gamma := \lambda_R/\lambda_D$. Table 7.1 summarises the results.

While the precise form of the cost curves depends, of course, on the input parameter values, a number of observations have been confirmed in a wide range of settings. First of all, it is not surprising that both s^* and $S^* - s^*$ are decreasing in γ. The higher the demand fraction that can be served by returns the less and the less often new items need to be ordered. Furthermore, total costs appear to be rather stable as long as γ is not close to one. Depending on parameter settings the cost function may moderately decrease or increase in γ. For return ratios close to one, though, total costs increase steeply. This is in accordance with earlier numerical results reported by Van der Laan and Salomon (1997). For an explanation, it is useful to consider the different cost components. Holding costs show a similar behaviour as total costs. For moderate values of γ additional stock caused by item returns appears to be roughly compensated for by a smaller reorder level whereas return ratios close to one entail excessive on–hand stock. Note that for high return ratios replenishment orders are rare and the on–hand inventory essentially behaves as an $M/M/1$ queue which is well known to 'explode' for high traffic intensity. Fixed order costs, on the other hand, tend to decrease towards zero along with the fraction of demand to be served from new items. The impact of this effect on total costs depends on the value of the cost parameters. Finally, backorder costs appear to be fairly stable.

To conclude, it seems that the return flow has a rather modest impact on the optimised expected inventory costs unless the return ratio is close to one, resulting in high on–hand inventory. It should be noted that this comparison does not include variable order costs. Incorporating per unit order costs c_v, the average variable order costs for a given return ratio equal $\lambda_D(1-\gamma)c_v$. To take the resource savings of product recovery into account, this linearly decreasing term should be added to the above cost components.

Standard material management tools often do not allow for explicit modelling of stochastic item inflows. Therefore, approximations are used to deal with return flows. We illustrate the performance of the following approaches to approximating the – possibly negative – net demand per period by a demand distribution with a nonnegative domain.

- *Neglect returns*: The return flow is not taken into account for determining the order policy. Hence, the parameter values $(s^*(0), S^*(0))$ are used for all values of γ.
- *Netting*: Returns are cancelled against demand. The net demand distribution is approximated by a first moment fit within a class of standard demand distributions. In the numerical example a Poisson distribution with mean $\lambda_D - \lambda_R$ is used.
- *Two moment approximation*: The net demand distribution is approximated by a two moment fit. The numerical example is based on a compound Poisson distribution with a discrete uniform distribution of the batchsize. (Note that the chosen distribution must be such that its mean goes to zero for a return ratio close to one, while the variance grows to infinity.)

Each of these approaches leads to a conventional inventory model for which optimal control parameters can be computed with standard methods. Figure 7.4 shows the resulting expected costs in the original model for each of these approaches. Not surprisingly, the deviation from the optimal policy increases with the return ratio. Moreover, the two moment fit appears to yield fairly good results for the entire range of return ratios. In contrast, a simple netting approach seems satisfactory only for relatively moderate return volumes. Otherwise, underestimating the system's variability entails a significant cost increase. Note that the variance of demand in the netting approximation decreases in γ while it increases in the actual system. 'Netting' implicitly assumes that returns occur exactly when they are needed (i.e. simultaneous with demand), which is not true in the exact model. Therefore, netting implies an underestimation of the required safety stock and hence

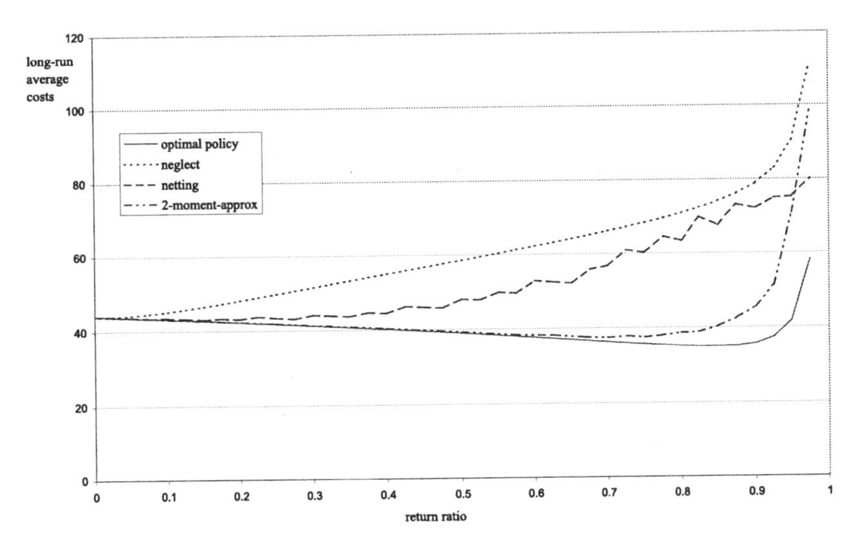

Fig. 7.4. Performance of alternative approximations

frequent backorders. Furthermore, it can be observed that ignoring returns in determining the control parameters may entail significant overstocks even for rather low return ratios and therefore is not a satisfactory approach.

Finally, we address the issue of information requirements. As discussed before, a high level of uncertainty is often stated as one of the main characteristics of Reverse Logistics environments. This is reflected in the above model in the growing coefficient of variation of the net demand as a function of the return ratio. However, the above analysis suggests that the exogenous character of the return flow concerns more than additional uncertainty only. Even if perfect information on the return flow were available it would be the potential mismatch between supply and demand that adds to the complexity in the inventory system. To assess the impact of both factors specifically we compare the original model with two alternative scenarios as follows.

To estimate the impact of uncertainty we consider a situation with perfect return information where return volumes are known beforehand for each period. Since an optimal order policy may have a complex structure in this case and can no longer be easily determined we use a myopic heuristic that only takes returns into account that arrive within the next procurement leadtime period. It should be noted that this is equivalent to assuming returns to undergo a recovery process having the same leadtime τ as procurement. We show in Section 7.6 that an (s, S)–order policy is still optimal in this case. Comparing this 'perfect information' scenario with the original situation gives an indication of the value of enhanced return information and hence of the potential savings that companies may realise through improved forecasting.

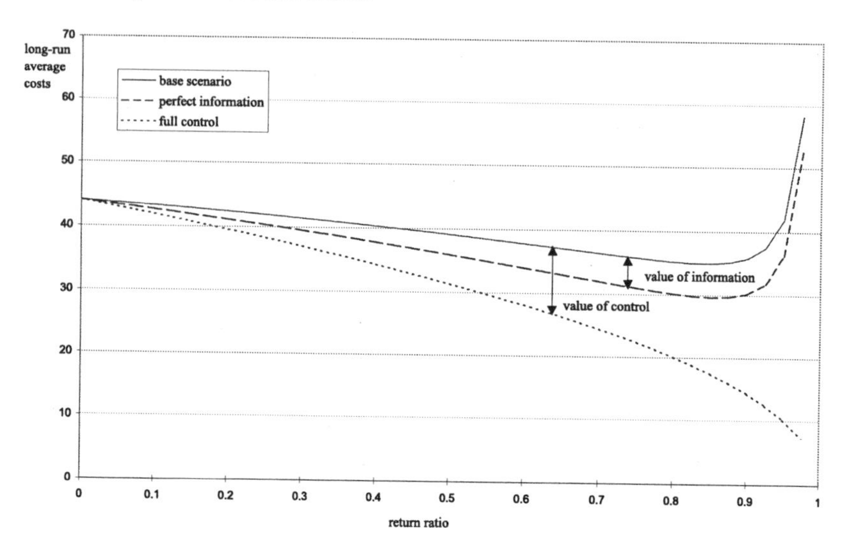

Fig. 7.5. Value of information and control

In a second scenario we assume that the timing of returns is no longer exogenously determined but can be actively controlled. While the overall return ratio cannot be changed, all returns can now be forced to coincide with a demand epoch. Note that this is equivalent to a standard (s, S)–model where the above 'netting' approach is exact. Comparing the costs of this scenario with the base case reflects the potential benefit of improving a company's control over its Reverse Logistics channel.

Figure 7.5 shows the minimum expected cost functions for the three scenarios, based on the above parameter settings. Since controlling the timing of returns includes the option of not interfering at all it comes as no surprise that the full control scenario yields the lowest costs in all cases. What is more important is the difference between the perfect information and the full control setting with respect to the asymptotic behaviour for large return ratios. Just as in the original situation, costs in the perfect information setting 'explode' as the return ratio approaches one. In contrast, controlling the timing of returns allows to reduce inventory costs towards zero. Hence, high stocklevels in situations with large return volumes may not be that much a matter of uncertainty but rather a matter of supply and demand imbalances. These results illustrate that companies should consider more than forecasting when looking for ways to efficiently integrate reverse goods flows into their materials management. Active return control, e.g., by means of specific take–back offers or closer relations with customers may entail significant additional benefits.

7.6 Extensions

We conclude our analysis by reconsidering the modelling assumptions and discussing possible extensions. As pointed out, the analysis presented for a discrete time setting can easily be carried over to a continuous time setting analogous with standard inventory models. To this end, assume demands and returns to be generated by a compound Poisson process. Letting $(p_i)_{i \in \mathbb{Z}}$ denote the batchsize distribution and $G(l)$ the expected variable costs incurred during an interoccurrence period starting with stock level l the continuous time model can be transformed into an equivalent discrete time model as in Section 7.2. We refer to Zheng and Federgruen (1991) for a discussion of more general compound renewal demand processes, which applies analogously to the return flow model.

Recovery process and leadtimes
Furthermore, a recovery process transforming returned items into serviceable items can be factored in, analogous with leadtime handling in traditional inventory models, under the condition that the recovery leadtime does not exceed the procurement leadtime. To be specific, we assume that each returned item undergoes some reprocessing. Therefore, returns during period n are added to the serviceable stock only at the beginning of period $n + \tau_R$ for some fixed recovery leadtime τ_R, which we assume to be smaller or equal to the procurement leadtime τ. Analogous with traditional inventory systems a replenishment decision in period n only affects the stock level after period $n + \tau$. Moreover, at that moment all outstanding orders and recovery work in progress at period n will have arrived. This motivates the definition of a modified inventory position at the beginning of period n (before ordering) as $\tilde{I}_n := Y_n$ + outstanding orders + recovery work in progress. Note that $\tilde{I}_n$ follows the same dynamics as the net stock in the original model. Moreover, let $\tilde{D}_n := \sum_{i=n}^{n+\tau-1} D_i^+ - \sum_{j=n}^{n+\tau-\tau_R} D_j^-$ denote the effective net demand during the procurement leadtime, distributed as an integer random variable $\tilde{D}$. Then we have $X_{n+\tau} = \tilde{I}_n + A_n - \tilde{D}_n$, and $\tilde{D}_n$ is independent of $\tilde{I}_n$. As in standard inventory models it is therefore sufficient to consider order policies depending on the (modified) inventory position. Moreover, the decision–relevant variable costs are $\tilde{G}(x) := \mathbb{E}[G(X_{n+\tau}) | \tilde{I}_n + A_n = x] = \mathbb{E}[G(x - \tilde{D})]$ which are convex in x. (Note that recovery costs and work–in–progress inventory are irrelevant in this context since they cannot be influenced.) This puts the extended model in the form introduced in Section 7.1.

Note that the recovery leadtime may also be stochastic as long as it does not exceed the procurement leadtime. In the case $\tau_R > \tau$ the above approach fails since $\tilde{I}_n$ and $\tilde{D}_n$ are no longer independent. In this case a more complex state variable is required distinguishing returns in individual periods and one may not expect an easily tractable optimal policy structure. As a heuristic one may modify $\tilde{I}_n$ such as to only take into account returns increasing the serviceable inventory within the next τ periods (similar with the leadtime modification proposed by Inderfurth and van der Laan, 1998). From a practical perspective the condition $\tau_R \leq \tau$ does not seem very restrictive. In many examples, recovering returned items is much faster, indeed, than producing or procuring new ones, in particular if the original product structure is preserved. In the example of IBM, procurement leadtimes for new spare parts often exceed a month whereas dismantling and testing parts from returned used machines is carried out internally within a week. A similar relation has been pointed out for Kodak's single–use camera production (see Toktay et al., 1999).

Disposal
A more serious limitation of our model concerns the absence of a disposal option. The numerical results in the previous section clearly indicate that disposal may be required to avoid excessive stock levels, in particular in the case of high return rates. In most real–life examples some disposal option can be expected to be available, possibly against additional costs. Including disposal in our model means allowing replenishment order sizes a_n to be negative. When (procurement and recovery) leadtimes are zero the model is then equivalent to a cash–balancing model (compare Section 6.1). In this case a double–sided (s, S)–policy is known to be optimal under some additional technical assumptions (Constantinides, 1976).

However, leadtime handling becomes substantially more difficult in this case. Since disposal immediately affects the on–hand stock the inventory position will, in general, no longer be sufficient to capture the system's state for an optimal policy. Therefore, a multi–dimensional state space is required and a simple optimal policy structure cannot be expected. Inderfurth (1997) has shown that the situation becomes easier for the special case of equal leadtimes for recovery and procurement where returns may be disposed of upon arrival. In this case, all decisions again affect the same future period and a simple critical number policy is optimal. While this study excludes fixed order costs one may expect a similar result for that case. To sum up, the above discussion shows that some trade–off has to be made between leadtime and disposal modelling to assure optimality of a simple control policy in inventory models with returns. In the general case more detailed information (in the form of a more complex state space) is required for mathematically optimal decisions. In practical applications one may therefore resort to more simple heuristics. In our model, controlling disposal by a critical upper bound on the on–hand stock may be a natural choice.

Independence of demand and returns
As discussed in Chapter 6, the relation between demand and return is one of the discriminating factors between the different inventory models that have been proposed for Reverse Logistics. It is easy to see that the proofs in the previous sections largely rely on the independence between both processes. This assumption appears to be justified in applications as, e.g., IBM's spare parts dismantling where there is no direct causal relation between demand

and returns. However, in other cases it may be more natural to assume returns to follow demand with a certain time lag. Specifically, several authors have assumed an exponentially distributed delay time (see Chapter 6). Our model can be extended to this special case as follows.

Assume that each item sold is eventually returned with a probability $p > 0$, whereas it is lost with probability $1 - p$. Items that will return have an exponentially distributed sojourn time in the market, sojourn times of different items are independent and include possible time on backorder. This model can be described as a two–dimensional Markov process (I_n, M_n) where I_n denotes the inventory position as before and M_n denotes the number of items in the market. By comparing sample–paths starting in states $(i, m+1)$ and $(i + 1, m)$ one can show that an optimal policy only depends on the state variable $I_n + M_n$. Moreover, the stationary distributions of $(I_n)_{n \in \mathbb{N}}$ and $(M_n)_{n \in \mathbb{N}}$ turn out to be independent so that the average cost function can again be written in the form of relation (7.12) with an appropriately chosen function $H(i)$. Hence, the above results for determining an optimal policy can be applied to this extended model, too.

However, it should be noted that the process M_n may not be directly observable in many practical situations. Therefore, the above exponential model should rather be seen as an easily tractable approximation. The approach relies crucially on the memoryless–property of the exponential distribution. For other market–sojourn time distributions far more complex state information is required for a mathematically optimal policy. In this case, assuming independence between demand and returns essentially means to ignore some part of the information that is available on future returns. The impact of this omission very much depends on the specific context. In general, one may expect demand information to be important for returns forecasting in cases of highly irregular demand (e.g. due to seasonal peaks) and rather short market sojourn times. For example, Toktay et al. (1999) report on an average market sojourn time of 8 weeks for Kodak's single–use cameras. Similarly, Goh and Varaprasad (1986) show that about 70% of reusable softdrink bottles are returned within the same month and more than 95% within two months. However, in other cases the time in the market is much longer. Durable products such as electronics equipment or cars are typically only returned after several years. In this case, variability in the market sojourn time is large compared to the inventory system's average time between ordering. Therefore, exploiting the correlation between demand and returns may primarily be useful for updating the return rate during a product lifecycle rather than for controlling individual orders. To sum up, we conclude that assuming independent demand and returns is at least a good approximation in many cases, which yields a simple and easily implementable control policy.

As pointed out earlier, the analysis in this chapter has focused on the end–item level. The next chapter considers the logistics design of the recovery channel in more detail.

8. Impact of Multiple Sources

8.1 Tradeoffs Between Recovery and Procurement

Another major characteristic of inventory systems in a Reverse Logistics context as identified in Chapter 6 concerns the presence of multiple, alternative supply sources, namely recovery of used products versus procurement of new ones. In the previous chapter we assumed that the goods return flow directly affects the serviceable inventory and cannot be influenced. Therefore, procurement orders have, in fact, been the only means to control the system. Let us now turn to a more detailed picture of the inbound channel where the recovery of returned products is also a decision variable (see Figure 6.1). To this end, we assume that returned products are collected in a distinct inventory upon arrival. The serviceable stock can then be replenished alternatively by means of procurement or by processing recoverable stock. In addition to lotsizing and safety stock considerations the choice between both sources then becomes an important issue in achieving efficient system performance.

As discussed in Chapter 6 multiple supply sources in traditional inventory systems mainly concern regular versus emergency deliveries. A tradeoff is made between a higher procurement price and a leadtime reduction. However, the reasoning in a Reverse Logistics context appears to be a different one. Rather than a leadtime reduction, it is the restricted availability of the cheaper (recovery) source that calls for an alternative supply source in this case. In contrast, leadtimes are longer for procurement than for recovery in many cases (compare Section 7.6).

It is clear from the discussion in the previous chapter that one may not expect that mathematically optimal solutions to this problem have a simple structure, in general. In fact, we have seen in Section 7.6 that the problem often becomes intractable even under additional simplifying assumptions. When explicitly distinguishing the above two types of inventories an optimal policy structure is only known for the case of negligible fixed costs and equal leadtimes for both sources (Simpson, 1978; Inderfurth, 1997; see also Section 6.3). From a practical point of view it therefore seems more important to come to a good understanding of the major tradeoffs governing the above situation in order to derive appropriate heuristic decision rules.

Assuming recovery to be cheaper and to have the shorter leadtime of both sources, the considerations driving the procurement decision appear to be very similar to the analysis in the previous chapter. Expecting all available recoverable items to be used before resorting to additional new supply, procurement decisions should depend on the aggregated stock rather than on the two individual inventory levels. Note that this brings us back to the model of Chapter 7. Therefore, one may expect a critical number policy based on the aggregated stock to be reasonable for procurement decisions here. In the next section we illustrate additional effects due to fixed recovery costs. Should the recovery leadtime, indeed, exceed the procurement leadtime substantially one may think of procurement as a kind of emergency supply. In this case the serviceable inventory level may become a more appropriate trigger for order decisions than the aggregated stock. Traditional two–source models may then provide a reasonable approximation. We repeat, however, that a recovery leadtime excess does not seem typical of many Reverse Logistics environments and therefore do not address this case in the remainder of our analysis.

Control of the recovery process deserves some more detailed considerations. Drivers for building up a distinct inventory of returned products rather than recovering them immediately may be twofold. First, fixed setup costs of the recovery process may call for a sufficiently large lotsize. Second, holding costs for the recoverable inventory may be lower than for serviceables due to lower inventory valuation. On the other hand, delaying recovery of returned items implies a loss of responsiveness to demand and may therefore require higher safety stocks. In addition, stock volumes may be larger for this strategy if only a certain fraction of the returned products can actually be recovered, e.g., due to quality limitations which are only identified during the recovery process.

Van der Laan (1997) provides a detailed analysis of different strategies for controlling the recoverable stock. In particular, push versus pull driven strategies are compared. In the first case the recoverable stock is flushed whenever it exceeds a certain trigger level. The recoverable inventory is mainly motivated by economies of scale in the recovery process and therefore has a lotsizing stock character primarily. Note that the model in Chapter 7 can be interpreted as a recovery–push model with a critical lotsize of one (see Section 7.6). In the second case the recovery process is demand driven and is initiated whenever the serviceable inventory drops below a certain trigger level (under the additional condition that the available recoverable stock is sufficient for a certain minimum recovery lotsize). This approach is closer to a 'just in time' philosophy where activities are postponed until actually required. It should be noted that these push–versus–pull considerations are very similar to issues discussed for logistics network design in Section 5.1. They again reflect the role of Reverse Logistics as a link between the market forces on the supply and demand side. In contrast with conventional supply chain philosophy Reverse Logistics processes are not entirely demand driven. Instead, exogenous factors on the demand and supply side need to be matched. Delineating the impact of both drivers gives rise to the push–pull issues discussed above. As for the networks in Chapters 4 and 5 one cannot conclude in general that either a push or a pull strategy is superior in controlling the recovery process in the above inventory system. Based on the above discussion one may expect that a pull approach involves higher total inventories but a lower serviceable stock. The net result of this tradeoff in a concrete example depends on specific parameter values. In particular, the holding cost differences between recoverable and serviceable stock play an important role.

In this context it is worth mentioning that some authors have recently discussed the appropriate determination of inventory carrying costs in a Reverse Logistics context (see Corbey et al., 1999; Teunter et al., 2000). They have pointed out that linear holding costs in a Reverse Logistics setting cannot always be interpreted as a first order approximation of opportunity costs, in contrast with what is common practice in traditional inventory models. Therefore, the authors argue that the use of linear holding costs may be questionable from a discounted cash–flow perspective. While we do not enter this discussion in detail here we note that holding costs include other aspects such as storage capacity and the risk of obsolescence. Moreover, the proficiency of different cost criteria, such as average costs versus net present value, does not appear to be an issue that is specific to Reverse Logistics. Therefore, we follow common practice and assume a linear cost charged for carrying inventory.

8.2 The Capacity Aspect of Product Returns

In the previous section we have argued that a critical number policy based on the aggregated stock level may often be appropriate for controlling procurement orders in a Reverse Logistics environment. Some additional observations are worth mentioning that concern the effect of fixed costs of the recovery process. In particular, it may be surprising to note that an optimal procurement policy is not necessarily monotonous in the recoverable stock level. In contrast, an increase in the number of recoverable products on hand may justify a larger procurement volume in order to exploit a future recovery opportunity.

To make things concrete consider the following simple example. Consider the framework of Figure 6.1 in a stationary, deterministic context ignoring the disposal option. We address a periodic review system and assume the following sequence of events. At the beginning of each period both the serviceable and recoverable inventory level is reviewed. Then procurement and recovery orders are placed and delivered immediately. Subsequently, demand and returns are realised. End of period stock gives rise to linear holding costs, backorders are not allowed. The average order and inventory costs are to be minimised. We use the following notation.

d_n	$=$	demand in period n;
r_n	$=$	returns in period n;
I_n^s	$=$	serviceable stock at the beginning of period n before ordering;
I_n^r	$=$	recoverable stock at the beginning of period n before ordering;
Q_n^p	$=$	procurement order size in period n;
Q_n^r	$=$	recovery order size in period n;
K^p	$=$	fixed procurement costs per order;
K^r	$=$	fixed recovery costs per order;
h^s	$=$	linear holding costs for end of period serviceable inventory;
h^r	$=$	linear holding costs for end of period recoverable inventory.

The system dynamics are then given by $I_{n+1}^s = I_n^s + Q_n^p + Q_n^r - d_n$ and by $I_{n+1}^r = I_n^r - Q_n^r + r^n$.

In the following example assume $d_n = 4$ and $r_n = 1$ for all n and set $K^s = K^r = 2$ and $h^s = h^r = 1$. Intuitively, one may expect that it is optimal in each period either to recover everything or to raise the the serviceable inventory to $d_n = 4$ by procurement. However, the following policy turns out to be optimal instead (omitting the period index for ease of notation).

$$(Q^p, Q^r)(I^s, I^r) = \begin{cases} (0,\ 0) \text{ if } I^s \geq 4 \\ (0, I^r) \text{ if } I^s + I^r \geq 4,\ I^s < 4 \\ (4 - I^s,\ 0) \text{ if } I^s + I^r < 4,\ I^r \neq 2 \\ (5 - I^s,\ 0) \text{ if } I^s + I^r < 4,\ I^r = 2 \end{cases} \quad (8.1)$$

See Figure 8.1 for a graphical illustration. Considering the transitions shows this policy to have a unique recurrent set consisting of the states $(0, 1), (0, 2)$ and $(1, 3)$ (marked in grey). Moreover, the average costs amount to 13/3 per period.

It should be noted that the above policy is non–monotonous in the columns $I^s = 0$ and $I^s = 1$. In both cases, one more item than we may intuitively expect is procured for $I^r = 2$. The explanation of this effect is that the additionally procured item allows us to satisfy demand in the following period by means of recovery only. Otherwise an additional procurement order would be required. If, on the other hand, the initial recoverable inventory is

recoverable stock \ serviceable stock	0	1	2	3	4	5
1	***(4,0)***	(3,0)	(2,0)	(0,1)		
2	***(5,0)***	(4,0)	(0,2)			
3	(4,0)	***(0,3)***			no action	
4	(0,4)					
5		recover all				

Fig. 8.1. Non–monotonous optimal policy

lower than 2 then the additional serviceable stock required to render recovery sufficient in the next period is too large to pay off.

It is worth noting that a similar effect is known for inventory systems with a capacity restriction on the replenishment size. Also in that case a higher stock level may call for a larger order–size if this allows for a better capacity utilisation in subsequent periods. More specifically, Chen and Lambrecht (1996) show an optimal policy to have the following structure: If the inventory level is below some critical bound an order of full capacity is placed; if the inventory is above some other bound no order is placed; between both bounds the optimal order size does not necessarily decrease monotonously in the inventory level but may have cyclic patterns depending on the relation between the EOQ–lotsize and the capacity limit. This parallel with the above Reverse Logistics example emphasizes once more the capacity aspect of product returns. The available recoverable stock can be interpreted as a (variable) capacity limit on the recovery process. We recall that we have observed a similar effect in the network design models in Chapter 5.

We conclude this section by showing that the proposed policy is indeed optimal in the above example. Optimality can be verified by solving the optimality equation of the corresponding dynamic programming problem. More easily, however, the following considerations show that this policy is even the unique (up to transient states) optimal stationary policy. To this end recall that the above policy yields an average cost of $13/3 = 4\ 1/3$. To find candidate policies achieving the same or lower average costs consider a lower bound $\underline{c}(.)$ on the direct costs in state (I^s, I^r). We recall that holding costs concern the inventory at the end of a period, after demand and return realisation. Since this is equal to the inventory at the beginning of the subsequent period (before ordering) the average costs remain unchanged when assigning the holding costs to the starting inventory in each period. Under this re–allocation the costs incurred in state (I^s, I^r) are at least $I^s + I^r$. Moreover, in any state with $I^s < 4$ at least one setup is required entailing additional costs of 2. This yields the lower bounds $\underline{c}(I^s, I^r)$ as indicated in Table 8.1.

Note that there are at maximum three states yielding costs that do not exceed the above average of 4 1/3, namely $\mathcal{S} := \{(0,1);(0,2);(1,1)\}$. Further-

Table 8.1. Lower bounds on one period costs

recoverable stock	serviceable stock 0	1	2	3	4	5
1	*3*	4	5	6	5	6
2	*4*	5	6	7	6	7
3	5	*6*	7	8	7	8
4	6	7	8	9	8	9
5	7	8	9	10	9	10

more, note that in this deterministic setting each recurrent state is visited with the same frequency. Therefore, a recurrent cycle can include at most one state with costs of 6 or two states with costs of 5 in addition to $\mathcal{S}$ without exceeding the average cost of 4 1/3. Finally, taking into account that all states in $\mathcal{S}$ require a procurement setup and that $I_{n+1}^r = I_n^r$ unless an additional (recovery) setup is paid shows that $(0,1) \to (0,2) \to (1,3)$ is indeed the only possible cycle leading to average costs of at most 4 1/3.

Conclusions of Part III

The past three chapters have addressed inventory management in a Reverse Logistics context. Several business examples have illustrated the issues that companies face as they incorporate flows of secondary resources in their material management. Comparing these scenarios with traditional inventory management situations, two main aspects appear to give rise to additional complexity. On the one hand, one needs to incorporate exogenous inbound goods flows, which may raise stocklevels. On the other hand, multiple alternative supply sources need to be coordinated, namely the recovery of returned goods and the procurement of new ones. A special class of examples of recoverable inventory management that has been around for a long time concerns rotable spare parts systems. However, more recent examples involving end–of–use product returns appear to be significantly different in that inbound and outbound flows do not form a closed loop where every return triggers an immediate demand for a replacement. Therefore, efficient inventory management in a Reverse Logistics environment has been shown to require novel approaches.

In general, inbound goods flows complicate the mathematical analysis of inventory systems due to a loss of monotonicity and a lack of an upper bound on the stocklevel. However, it has been shown in Chapter 7 that inbound goods flows can be incorporated in the structure of standard inventory control models under certain conditions. In particular, demand and returns have been assumed to be independent and disposal has not been included. In this case, the stocklevel can be decomposed into a part that behaves as in a conventional inventory system and a part that is independent of replenishment orders. As a consequence, the return flow model can be transformed into an equivalent standard inventory model without returns. Standard algorithms can then be applied to determine optimal parameters for controlling replenishment orders. Moreover, applying general theory on Markov decision processes an (s, S)–order policy has been shown to be optimal under these conditions.

The presented approach relies crucially on the independence of demand and returns. While in many applications returns are dependent on previous demand we have argued that the scale of this time lag is often much larger than the planning horizon of individual order decisions. Therefore, assuming demand and returns to be independent in the short run appears to be justified in many cases. Rather than for determining individual replenishment decisions the correlation between demand and return may be exploited for updating the system's parameters in the course of a product lifecycle.

In many other, more general inventory models with Reverse Logistics one may not hope for an easily tractable optimal policy structure. However, we have seen that the above model may serve as a basis for developing reasonable heuristics for these cases. Specifically, approaches have been discussed for factoring in a disposal option and a refined modelling of the recovery channel involving an additional stockpoint. Moreover, external drivers influencing supply and demand have been shown to give rise to tradeoffs between push and pull controlled processes. In the previous part of this monograph it has been discussed in the context of logistics network design that the available volume of recoverable products can be interpreted as a capacity constraint on the recovery process. It is worth mentioning that this perspective is supported by an analysis of the corresponding inventory economics. It has been shown that inventory models including a Reverse Logistics source may result in similar mathematical structures as capacitated inventory control models.

Numerical examples have shown that the volumes of Reverse Logistics flows tend to have a rather limited impact on the overall inventory costs if properly taken into account in the replenishment decisions. The value recovered from returned products may therefore be expected to outweigh a possible increase in inventory costs in many cases. Only for very high return ratios disposing of recoverable products may be required in order to avoid excessive stocklevels. However, to assure good performance it is crucial to take Reverse Logistics flows explicitly into account in inventory management. Simplistic approaches, such as ignoring product returns until they are actually available or netting demand and returns, may entail a significant cost burden. In order to approximate the behaviour of the inventory system appropriately both the volume and the variability of Reverse Logistics flows need to be taken into account.

Finally, recall that Reverse Logistics flows impose new challenges on inventory management due to a growing level of uncertainty. Enhanced information on future product returns, e.g., by means of improved forecasting or monitoring may therefore help to reduce inventory costs. However, even larger savings may be realised by extending a company's control over its inbound goods flows. Rather than reacting to whatever Reverse Logistics flows arise companies should therefore look for ways to actively manage used goods flows as a resource to satisfy customer demand.

Part IV

Reverse Logistics: Lessons Learned

9. Integration of Product Recovery into Spare Parts Management at IBM

In this chapter we return to the business example of IBM. For the general context of IBM's Reverse Logistics activities we refer to Chapter 2. Now focus is on the integration of returned used equipment into the spare parts management more specifically. Applying conclusions from the previous chapters we propose a systematic logistics concept for this channel. The material presented in this chapter summarises the results of a joint study in co–operation with IBM's spare parts logistics division in Amsterdam.

Section 9.1 below describes the initial situation and discusses deficiencies concerning the way used equipment dismantling is dealt with at present. Section 9.2 delineates process alternatives and formulates a number of concrete decision problems. In Section 9.3 performance of the different options is compared quantitatively. To this end, a small simulation model is developed. Section 9.4 condenses our conclusions into a number of recommendations.

9.1 The Current Dismantling Process

As discussed in Chapter 2 IBM manages an extensive spare parts network to support its service activities. Figure 9.1 sketches the network structure for Europe, the Middle East, and Africa (EMEA). Similar networks cover America, Asia, and the Pacific region. The stock points for spare parts form a divergent multi-echelon structure where the allocation of each part depends on part value and service constraints. A central buffer (CB) in Amsterdam feeds lower level stock points, such as country stock rooms (CSR) and branch offices (BO). In addition to regular new buy, supply sources for the CB include manufacturing surplus, surplus from other geographical regions, and brokerage. Moreover, as explained in Chapter 2, for many parts a repair process is available. In that case, defective parts from the customer base are returned upwards in the network. They are kept in a distinct stock marked as 'available for repair' (AFR) and are repaired as needed. Parts from all of these supply sources must meet a common quality standard indicated as certified service parts (CSP).

The CB is controlled according to an MRP–policy. Based on forecasts, future orders are planned for which supply sources are selected on price and

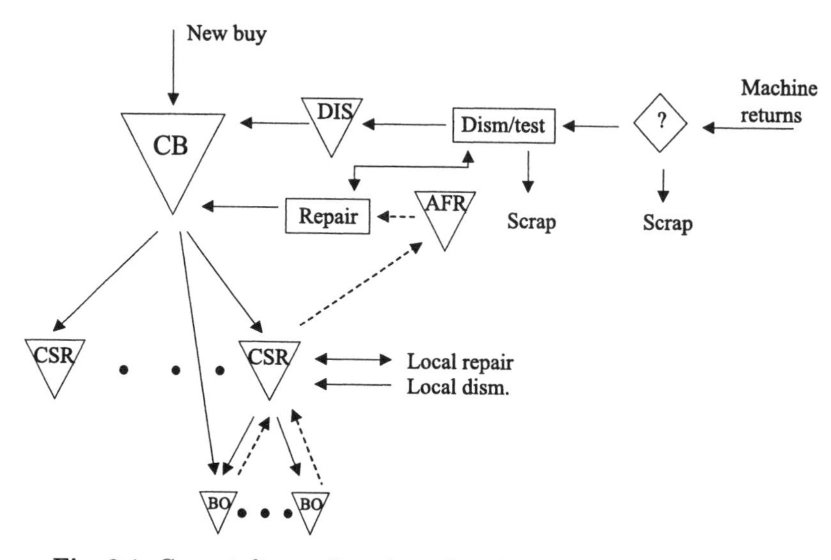

Fig. 9.1. Current dismantling channel in the IBM spare parts network

availability. Planned orders are reviewed in a four-weekly cycle and may be adjusted until they arrive within a time fence where orders are 'frozen'.

As discussed in Chapters 2 and 6, dismantling of returned used machines has recently emerged as a complementary source of spare parts. In addition to refurbishment of returned equipment for extended use, extracting parts for use as spare parts is an opportunity to recover value from used machines. This option is the more attractive since the economic and technical lifespan of individual parts often substantially exceeds that of the complete machine. In order to guarantee quality standards, parts from dismantled machines need to be carefully tested and possibly repaired before they can be reused. Typically, these latter costs are significantly lower than those for buying new parts. In addition, availability of new parts may be limited as a consequence of short innovation cycles. In general, maintaining a manufacturing process for spare parts only is not economical. In this case, dismantling may provide a valuable source for parts during the remainder of the service horizon. The cost differences between dismantling and the traditional repair channel are less salient. Yet dismantling may have some cost advantage since parts from used equipment are not necessarily 'defective', in contrast with exchange parts returned from the field. Therefore, one may expect higher yields and possibly allow for simpler test processes to meet quality standards. To sum up, dismantling is an opportunity for reducing service parts procurement costs by recovering value from returned machines.

Dismantling has initially been set up on a small scale for retrieving high value parts from end-of-lease equipment. Making use of dismantling as a source of parts has largely been a manual process that is not supported by the overall planning system. As return volumes of end–of–life equipment are increasing, due to commercial and environmental considerations (see Chapter 2), limitations of this approach are becoming apparent.

More specifically, the current dismantling process feeding the CB in Amsterdam is as follows. Returned machines that cannot be reused as a whole are offered for dismantling. Based on experience and on forecast demand, engineers determine which parts to dismantle from a given machine. Physically, dismantled parts are then added to the AFR stock so as to undergo the same process as repair parts. However, administratively adding dismantling parts to the AFR stock is not possible. Therefore, dismantling parts are kept track of as a distinct type of stock (DIS). Since this stock is not visible to the service parts planning and ordering system no automatic orders on the DIS stock are generated. Consequently, dismantling parts need to be pushed manually into the CSP stock by cancelling orders on other, more expensive sources and substituting them with supply from the DIS stock. Since this artificial distinction between physical and administrative handling cannot be made with external repair vendors, dismantling is restricted to parts for which test- and repair-processes are available in-house. In addition to supplying the CB, dismantling is also carried out locally, feeding individual CSRs. In these cases dismantling parts often undergo a simplified test process only, rather than the complete process of the repair channel. However, for reasons of quality standards parts from such local dismantling processes are restricted to use within the local organisation and may not be propagated through the network.

The above description shows that dismantling is not integrated as a regular source in the spare parts planning in the current situation. It is easy to see that this state of affairs involves a number of shortcomings. First, there is no concise dismantling rule determining which parts to extract from a given machine. The current, purely experience–based dismantling decision bears the risk of missing dismantling opportunities (that would reduce parts procurement costs) on the one hand, and of dismantling parts that are not used on the other hand. Second, dismantling is not actively used for procuring parts. Considerable effort is required to work around system limitations and to push dismantling parts into the network. This entails the risk of not using parts that are available from dismantling and a waste of resources in artificial activities. Third, control of other sources does not take dismantling into account. This may result in lost opportunities for saving parts costs since orders are placed on more expensive sources and parts from dismantling are no longer needed when they become available. Finally, the technical processes for making dismantling parts ready for reuse are not defined in a consistent way. Uncritical consideration of dismantling parts as being equivalent to repair parts may be overly conservative and result in a loss of savings opportunities for parts that are still functional. To overcome these drawbacks a systematic process is sought for integrating dismantling as a regular element into the parts planning.

9.2 Logistics Alternatives for Integrating Dismantling

It should be noted that the above situation can be embedded in the framework developed in Section 6. Dismantling corresponds with the 'recovery' channel while the regular 'procurement' channel encompasses several sources here, in particular new buy and repair. Moreover, for each part the 'serviceable stock' represents the entire network stock, which is replenished via the central buffer. To address the goal of developing an appropriate logistics concept for parts dismantling systematically we can distinguish process design and control issues. The prior concern the logistics structure of the dismantling process. The latter can be further divided into the need for a dismantling decision rule and the co–ordination with other supply sources. We address each of these issues in a separate subsection below.

9.2.1 Design of the Dismantling Channel

First of all, the processing steps need to be clearly defined which a part from a returned machine has to undergo before being reused. This refers to both technical steps and their logistics linkage. In general, processing includes the actual dismantling, upgrading and testing, and possibly repair. Both from a quality and a cost perspective, testing is the main step in this channel.

The engineering division is responsible specifying for the test processes that are needed to bring a dismantling part to a certified quality level compatible with service parts quality requirements. In this context the relation between the dismantling channel and the traditional repair channel is of major importance. In fact, 'repair' to a large extent consists of testing. The same procedure may therefore be applied to dismantled parts. However, as discussed above, parts from returned equipment can, in general, be expected to be in a better condition than returned exchange parts. Therefore, testing dismantled parts may be an option even if there is no repair process. Moreover, in some cases returned machines may be tested as a whole rather than on an individual part level. These so called 'box tests' are mainly applicable to 'simple' parts such as PC components.

From a logistics angle one needs to define the parts flow through the different processing steps and their co–ordination. The most simple approach is to add dismantled parts to the AFR stock. In this case, integration of dismantling is mainly a question of overcoming limitations of the current information and control system. This solution seems appropriate if quality differences between dismantled parts and returned exchange parts are small or if the costs of the actual 'repair' steps are small relative to the test costs. Otherwise, dismantled parts should be distinguished from the AFR stock even if the physical processes for both parts categories are partly the same.

In the case of a distinct dismantling channel one needs to decide upon inventory buffers decoupling the individual processing steps. As testing is the main value–adding activity in this channel, postponement of this step appears to be worth considering. In contrast, since costs for dismantling tend to be small and since normally only a few parts are retrieved from a given machine, stocking entire machines to postpone dismantling does not seem to make sense. Another reason for immediate dismantling concerns uncertainty with respect to the machine content. It turns out that the components that are actually found in a returned machine may differ significantly from what would be expected according to specifications. Possible explanations for this phenomenon include product model variations, a lack of administration concerning product modifications during use, and a lack of control concerning changes made by the customer. Therefore, reliable information on the content of a returned machine is only known after inspection. Dismantling while the machine is handled requires little additional effort.

For the test activities postponement considerations may be more relevant. Analogous with the repair channel one may keep an inventory of dismantled parts, on which (test–)orders are placed as needed. Alternatively, one may test dismantled parts immediately and push them into the CSP stock. Note that this brings us back to the considerations on the push versus pull control of the recovery process as discussed in Chapter 8. As explained earlier the pull approach has the advantage of postponing the expenses for testing and reducing the risk of unnecessary activities, which is typically reflected in a smaller holding cost for the intermediate inventory. On the other hand, a push approach can be expected to lead to smaller total inventories by reducing throughput times and eliminating defective parts. We compare both approaches numerically in Section 9.3.

In the case of a box test parts cannot be removed from the returned machine before having been tested. Therefore, postponing the test operation until parts are needed implies stocking entire machines in this case. On the other hand, the above reasoning suggests that parts should be dismantled directly once testing has been completed. Therefore, essentially the same alternative channel structures apply as above. While in the above setting the 'recovery process' in terms of Figure 6.1 concerns testing it now includes both testing and dismantling.

9.2.2 Dismantling Decision Rule

In addition to the process design, operational control rules need to be specified in order to integrate dismantling into the parts planning. This includes determining which parts to dismantle from a specific machine that has been returned. Since dismantling opportunities, in general, do not occur exactly at times when parts are needed a tradeoff has to be made between building up additional inventory and potential savings in procurement costs. More specifically, comparing the direct cost implications of dismantling with those of disposing a certain part suggests dismantling those parts for which the expected holding costs incurred until the moment of use are less than the cost difference between dismantling and testing on the one hand and procurement costs for alternative supply plus disposal costs on the other hand. This rule, which boils down to specifying a critical dismantling horizon for each part, has been proposed by Inderfurth and Jensen (1998) for a deterministic MRP–setting with product return flows. We remark that the 'expected moment of use' should also take into account commercial and technological developments such as potential parts design changes. Therefore, one may consider upper bounds on the dismantling horizon, independent of the cost parameters. Furthermore, availability of an appropriate test process should be reflected in the 'test costs'. It is worth pointing out that dismantling decisions for different parts are fairly independent since testing is the dominant cost factor. Only in the case of box testing should an integrated measure on the individual parts be used to decide whether or not to test a given box.

It should be noted that the above decision rule is somewhat heuristic in that it only takes into account the given part and ignores effects on future deliveries. Moreover, it is not immediately clear which 'alternative supply source' to consider to estimate the procurement cost savings appropriately. We analyse the impact of the critical dismantling horizon numerically in the next section. Finally, we remark that the above rule is equivalent to the disposal policy discussed in Chapter 8. Given a demand forecast a critical time span until use is equivalent to a critical level of serviceable plus recoverable stock.

9.2.3 Co–ordination with Other Sources

Establishing dismantling as a regular parts source also has an impact on the control of the other procurement sources. For good system performance, ordering decisions on all sources need to be co–ordinated. Since the dismantling channel involves lower costs and shorter leadtimes than new buy and repair it is straightforward that order triggers for these two sources should take into account both the serviceable stock and a possible inventory of dismantled parts. A more complex issue is the question of how to deal with expected future dismantling opportunities that are not known yet. In Section 7.5 we have seen that simplistic approaches such as ignoring future product returns or netting may be highly suboptimal. More specifically, in applying the previous results to the present context one may expect that ignoring future dismantling opportunities until they actually occur leads to excessive stock levels. On the other hand, relying on machine returns that are not yet known with certainty entails a higher risk of shortages since supply may vary and the dismantling volume may turn out lower than predicted.

In the model in Chapter 7 we have seen how stochastic recovery opportunities can be appropriately incorporated by computing a modified demand distribution. However, it should be noted that this approach requires the distribution of the return process to be known. This is not the case for the current situation at IBM. In contrast, little historical data is currently available on machine returns. Therefore, treating returns as an unknown factor may be the only option to start with. Gradually, more detailed data may then be collected to develop appropriate return forecasting schemes.

We recall in this context that machines available for dismantling stem from several sources, namely end–of–lease returns, trade–in equipment, and environmental take–back. Therefore, one may consider different forecasting approaches for each category. For example, one may expect that fairly detailed information can be obtained concerning the lease channel. Data on the installed base and on current lease contracts may provide a basis for estimating related product returns. Other return categories appear to be more difficult to forecast. Past sales and product life–cycle information may provide a rough indication of the expected level of returns. We emphasize anew that the machine return process differs substantially from returns of defective exchange parts. As pointed out before, machine returns are not directly correlated with spare parts demand, in contrast with defective parts returns. Therefore, the variability of the net demand tends to be much higher. Moreover, the time between issue and return of a spare part primarily depends on technical parameters such as failure rates, which can be estimated by quality control methods. In contrast, the market sojourn time of a machine depends on customer preferences in the first place and involves many factors that are difficult to assess, such as economic welfare and technological progress.

In the next section we investigate the performance of alternative approaches to dealing with future dismantling opportunities and assess the cost savings potential of improved forecasting information. Moreover, we analyse the impact of a more active return strategy that does not consider machine returns as a purely exogenous factor.

9.3 Performance of Alternative Planning Approaches

We choose a simulation approach to compare the performance of the different logistics alternatives for integrating dismantling into the spare parts planning discussed in the previous section. While the results of Chapters 7 and 8 permit a direct computation of some of the performance measures simulation allows for a greater flexibility in considering various modelling modifications. We use our analytic results as a point of reference for judging the empirical outcomes. In the next subsection the simulation model is presented. Subsequently we report on the main numerical results.

9.3.1 A Simulation Model

The structure of the simulation model is illustrated in Figure 9.2 for the case of individual parts tests. In the case of box tests the dismantling step

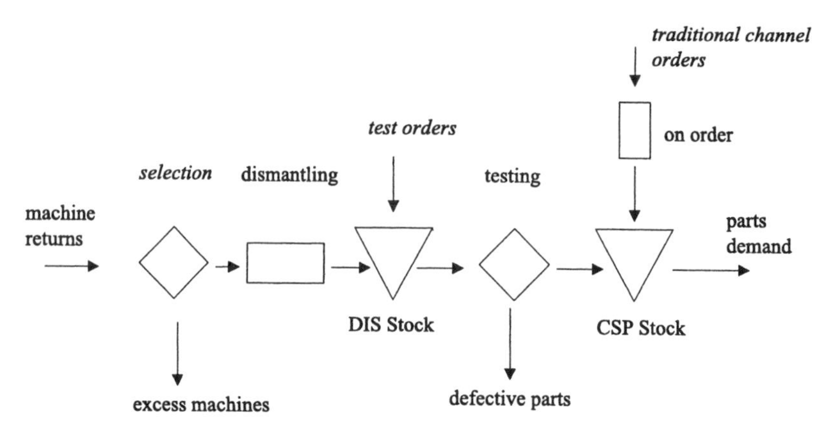

Fig. 9.2. Structure of the simulation model

is placed between testing and CSP stock. The first two columns of Table 9.1 summarise notation. The parts sample given in the remaining columns is addressed later. In the sequel we discuss the major modelling choices and assumptions.

- *Machine return process:* As discussed above, little historical data is currently available concerning machine returns. Experience suggests that machines arrive in rather small batches. We model machine returns as a stationary compound Poisson process with intensity λ_R and a batchsize that is uniformly distributed on $[1; b_u]$. While in reality the return volume can be expected to vary during the product life cycle we opt for a stationary approach in view of the lack of reliable data. Rather than trying to model the product life cycle explicitly, which requires the specification of numerous parameters we suggest evaluating the stationary model for different return flow levels. Furthermore, we consider the supply of individual parts. Recall that synergies in dismantling multiple parts are small. Specifically, we assume any arriving machine to be actually available for dismantling and to contain the part under consideration. Hence, dispositioning is beyond the scope of this model. We briefly address the impact of the dispositioning approach in the next section.
- *Selection:* Upon arrival of a returned machine a dismantling decision is taken based on the relation between current inventory level and expected demand. Following the argumentation in Section 9.2.2 a part is dismantled if its expected time in stock is lower than a certain critical level, which needs to be specified yet. Otherwise it is scrapped or salvaged externally for a (possibly negative) unit cost of c_s.
- *Dismantling:* The dismantling process is modelled with a fixed processing time τ_d and a unit cost c_d. As discussed earlier, economies of scale with respect to batchsizes appear not to play a major role at this stage. More-

over, we do not take capacity limitations into account since small arrival batches and a short processing time result in fairly small work in process queues.

- *DIS stock:* Following the discussion in Section 9.2.1 we consider the option of an intermediate stock point in the dismantling channel, postponing the test operation until required. Alternatively, items may be pushed through the test stage immediately. In the case of box testing the DIS stock concerns entire machines, otherwise it contains dismantled parts analogous with AFR stock in the repair channel. We follow the current AFR approach and value DIS stock at CSP value minus unit costs of the intermediate processes. DIS stock is then charged with a unit holding cost of h_r calculated as a fixed rate h times the stock value.
- *Test process:* The test process again involves a fixed leadtime τ_r and a unit cost c_r. A tested item is accepted with probability p_r and rejected otherwise. Test outcomes are independent of each other.
- *CSP stock:* As in the current system we value CSP stock at the weighted average cost (WAC) for the individual parts. Holding costs h_s are again determined as h times stock value per time.
- *Orders on traditional channel:* Since the focus of this analysis is on integrating dismantling as a new source into the current system we aggregate all existing sources in a single 'traditional channel' source with leadtime τ_n and variable cost c_n. As explained earlier, the current major sources concern new buy and repair. In the next section we comment on the robustness of our results with respect to modelling both sources explicitly.
- *Spare part demand:* Demand is modelled as a stationary Poisson process with rate λ_D. Recall that we prefer a stationary approach over a life cycle model as explained above. Demand in a stockout situation is taken care of via an emergency supply. Hence, there is no backlogging.

The above model encompasses three sets of decisions, namely concerning the selection step, control of the test process, and control of traditional channel orders. We compare the performance of the different strategies discussed in the previous section. More specifically, we consider the following options.

As in the actual current planning system, orders on the 'traditional' channel are controlled by a periodic review policy with a review period of one month. Since we neglect fixed order costs a single–parameter order–up–to policy is appropriate. The corresponding inventory position includes on–hand CSP stock plus outstanding orders plus expected yield of DIS stock and work in process. We consider three alternatives for setting the order–up–to level S_n, analogous with the approaches discussed in Section 7.5. First, we choose S_n such that 95% of the demand can be served directly from stock in the case of no dismantling. Parallel to Section 7.5 we refer to this policy as 'neglect' in the sequel. Second we reduce the order level by the expected dismantling contribution during a leadtime plus review period (denoted by 'netting'). Third, we adjust S_n such that the true servicelevel with dismantling equals 95%. As

Table 9.1. Parameter values of parts sample

Parameter	Description	**Part 1** *PC memory*	**Part 2** *CD-ROM drive*	**Part 3** *monitor*	**Part 4** *midrange HDA*	**Part 5** *high-end proc.card*
λ_D	part demand rate (per month)	200	150	100	10	2
λ_R	batch machine return rate	scenario dependent				
b_u	maximum return batchsize	10	5	5	5	2
p_r	test yield	0.8	0.75	0.9	0.8	0.7
τ_n	leadtime traditional channel	1.5	3	3	2	2
τ_d	leadtime dismantling	0.05	0.05	0.05	0.05	0.05
τ_r	leadtime testing / upgrading	0.15	0.1	0.15	0.2	0.2
c_n	unit order cost trad. channel	50	200	300	600	5000
c_d	unit cost dismantling	10	15	5	20	20
c_r	unit cost testing / upgrading	20	35	50	180	300
c_s	unit cost scrapping / salvage	-5	0	0	0	0
h	holding cost rate (per year)	0.2	0.2	0.2	0.2	0.2
h_n	unit holding cost serviceables	$h\times$weighted avg. proc. cost				
h_r	unit holding cost recoverables	$h_n - h(c_d + c_r)$				

this policy uses stochastic information on the return process it is denoted by 'forecasting'.

For the test process we consider a push and a pull approach. In the first case, each item is tested as soon as it is available so that there is no DIS stock. In the second case, test orders on the DIS stock are placed periodically, at the same review epochs as in the traditional channel. In each period as many items as available are released to the test stage as to bring the downstream inventory (CSP stock plus expected yield from test work in process) up to a target level S_r. We slightly adjust S_r depending on the policy for setting S_n. In the case of the conservative 'neglect' policy S_r is set to the expected demand during a test leadtime plus review period. For the more proactive 'netting' and 'forecasting' policies we increase S_r by a safety margin which is set according to the usual 'normal loss approximation' of the expected shortage per replenishment cycle (see, e.g., Silver et al., 1998).

Finally, the critical level H_d for the expected storage time of a dismantled part is set to the expected procurement cost savings divided by the unit holding costs, namely $[c_n + c_s - (c_d + c_r)/p_r] \;/\; [(c_d + c_r)h]$. Note that the holding costs in this expression are based on the dismantling channel costs

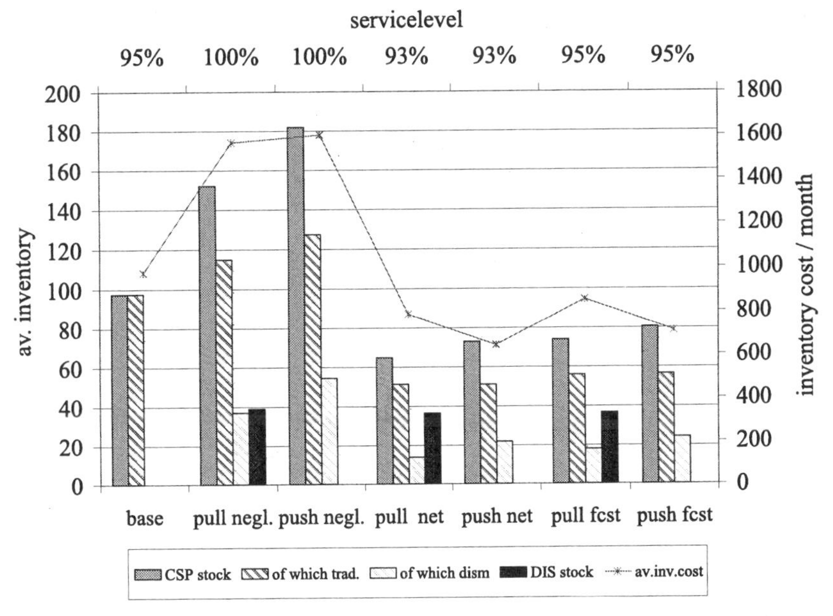

Fig. 9.3. Performance comparison for Part 1

rather than on the WAC since the latter depends itself on H_d. In addition, we restrict H_d to a maximum of two years, to take into account engineering design changes that limit parts reusability.

9.3.2 Numerical Results

We implemented the above model using the commercial software package Arena on a Pentium II PC. In this section we discuss our empirical results. We consider the five parts specified in Table 9.1. While the data presented does not coincide with any specific IBM parts the parameter settings reflect the relevant orders of magnitude. The monetary unit of all data is dollar and time is measured in months. Parts 1 through 3 concern relatively cheap parts with a high demand volume. Parts 1 (memory) and 2 (CD ROM drive) fall into the PC sector, Part 3 (monitor) is used in several product categories. Parts 4 (hard–disk assembly) and 5 (processor card) concern larger equipment having a smaller demand and being more expensive. Part 2 involves a 'box test' whereas all other parts are tested individually. The cost advantage of the dismantling channel as compared to the mix of other sources ranges from 25% (Part 1) to over 90% (Part 5). For each part the leadtime of the entire dismantling process, which is carried out in–house at IBM, is about one week, whereas leadtimes for traditional channels are in the order of several weeks to a few months.

Table 9.2. Simulation results

	Part 1						
	base	**pull negl.**	**push negl.**	**pull net**	**push net**	**pull fcst**	**push fcst**
S_n	477	477	477	357	339	379	355
S_r	–	240	–	240	–	240	–
H_d	–	15	15	15	15	15	15
Servicelevel	0.95	1.00	1.00	0.93	0.93	0.95	0.95
dism. cover	–	0.3	0.3	0.3	0.3	0.3	0.3
avg. inv. cost	975	1568	1602	783	642	852	709
CSP stock	97.5	152.0	182.0	65.0	73.0	74.0	80.6
from trad.	97.5	115.0	127.4	51.3	51.1	56.0	56.4
from dism.	–	37.0	54.6	13.7	21.9	18.0	24.2
DIS stock	–	38.9	–	36.5	–	36.6	–

	Part 2						
	base	**pull negl.**	**push negl.**	**pull net**	**push net**	**pull fcst**	**push fcst**
S_n	578	578	578	443	405	450	411
S_r	–	173	–	168	–	168	—
H_d	–	24	24	24	24	24	24
Servicelevel	0.95	1.00	1.00	0.94	0.94	0.95	0.95
dism. cover	–	0.3	0.3	0.3	0.3	0.3	0.3
avg. inv. cost	3000	5914	6290	2487	1761	2476	1841
CSP stock	75.0	174.0	202.9	72.8	56.8	76.2	59.4
from trad.	75.0	112.6	142.0	41.6	39.8	41.0	41.6
from dism.	–	61.4	60.9	31.2	17.0	35.2	17.8
DIS stock	–	41.0	–	29.8	–	29.9	–

	Part 3						
	base	**pull negl.**	**push negl.**	**pull net**	**push net**	**pull fcst**	**push fcst**
S_n	386	386	386	296	272	305	281
S_r	–	120	–	120	–	120	–
H_d	–	24	24	24	24	24	24
Servicelevel	0.95	1.00	1.00	0.93	0.94	0.95	0.95
dism. cover	–	0.3	0.3	0.3	0.3	0.3	0.3
avg. inv. cost	3102	5487	6147	2421	1875	2667	2070
CSP stock	51.7	113.9	135.7	49.8	41.4	54.1	45.7
from trad.	51.7	71.2	95.0	28.5	29.0	32.1	32.0
from dism.	–	42.7	40.7	21.3	12.4	22.0	13.7
DIS stock	–	23.6	–	16.5	–	16.6	–

	Part 4						
	base	**pull negl.**	**push negl.**	**pull net**	**push net**	**pull fcst**	**push fcst**
S_n	33	33	33	26	25	30	28
S_r	–	13	–	15	–	15	—
H_d	–	24	24	24	24	24	24
Servicelevel	0.95	0.97	0.99	0.90	0.91	0.95	0.95
dism. cover	–	0.3	0.3	0.3	0.3	0.3	0.3
avg. inv. cost	1092	1345	1408	850	814	1123	1024
CSP stock	9.1	11.2	15.2	7.9	8.8	10.2	11.0
from trad.	9.1	7.6	10.0	5.0	5.8	6.7	7.3
from dism.	–	3.6	5.2	2.9	3.0	3.5	3.7
DIS stock	–	4.9	–	2.5	–	3.1	–

	Part 5						
	base	pull negl.	push negl.	pull net	push net	pull fcst	push fcst
S_n	9	9	9	8	7	9	8
S_r	–	4	–	5	–	5	–
H_d	–	24	24	24	24	24	24
Servicelevel	0.95	0.95	0.98	0.94	0.92	0.95	0.95
dism. cover	–	0.3	0.3	0.3	0.3	0.3	0.3
avg. inv. cost	4200	4625	3795	3605	2616	4332	3189
CSP stock	4.2	3.9	5.7	3.8	4.0	4.3	4.8
from trad.	4.2	2.9	3.7	2.6	2.5	3.0	3.1
from dism.	–	1.0	2.0	1.2	1.5	1.3	1.7
DIS stock	–	2.4	–	1.4	–	1.8	–

Table 9.2 lists the simulation results for these parts for a potential dismantling volume of 30% of the demand. For each part a comparison is given of the six alternative policy combinations discussed above. Moreover, a base case without dismantling is added as a point of reference. Results for each case are based on 5 replications of a simulation during 5000 periods with a warm–up phase of 100 periods. For all statistics the 95% confidence intervals are in the order of one percent. Table 9.2 shows the values of the control parameter and performance indicators for each case. More specifically, we indicate the resulting service level (as the fraction of demand satisfied directly from stock), the actual fraction of demand covered by dismantling, the average inventory cost, and the average stock levels, including CSP stock and DIS stock. The former is further divided into the contribution of the traditional channel and the dismantling channel, respectively, which determines the CSP stock value. In addition, Figure 9.3 illustrates the results for Part 1 graphically.

We make the following observations. The largest performance differences are found between the alternative approaches for taking into account the expected future dismantling supply in setting the order up to level S_n of the traditional channel. As in Section 7.5 we see that neglecting future product returns results in excessive stocks whereas a 'netting' approach underestimates the risk of stockout and may lead to an unaccepted service level. The scale of the cost reduction from a conservative 'neglect' policy to an optimised forecasting based approach depends on the product return volume and ranges from about 10% for the 'slow mover' Part 5 to more than 60% for the high volume parts 1–3. We recall that the currently available data on product returns is fairly limited and may allow for little other options than neglecting dismantling inflow until it is has actually arrived. The cost difference between 'neglect' and 'forecasting' should therefore be seen as an indication of the savings potential of building up accurate product return data.

The impact of the additional DIS stock point appears to be rather limited: the performance differences between a pull and push control of the dismantling channel are fairly small. In line with our argumentation in Section 8.1 we find that a pull approach leads to a higher total stock volume whereas a push approach involves a higher CSP stock. The net effect of this tradeoff depends on the specific parameter settings. As a general tendency, we find that a pull policy tends to be slightly superior in combination with a 'neglect' approach whereas a forecasting–based policy may benefit from a dismantling push. The only exception to this rule is found for Part 5 where the extremely high procurement cost differences result in an advantage for the push policy in all cases.

Finally, we find that no reusable parts are disposed of in any of the cases considered. For all but Part 1 the cost–oriented maximum storage time for dismantled parts largely exceeds the technically determined limit of two years. Moreover, the actual throughput times turn out to be in the order of a few weeks in all cases, which is far below the critical limit H_d. Therefore, all dismantling policies make use of the maximum dismantling potential. In particular, all dismantling policies result in the same supply mix and hence in the same overall procurement costs. (In contrast, note that the procurement costs in the 'base' case are higher and that this cost difference, which amounts to $\lambda_R(b_u+1)(p_r(c_n+c_s)-c_d-c_r-c_s)/2$, is not included in the results in Table 9.2.) We conclude that the above results are very robust with respect to changes in H_d and hence that determining the maximum allowable storage time more exactly has little effect.

The above results are confirmed in a broader setting. Table 9.3 summarises the simulation outcomes for product return volumes allowing dismantling to cover 10%, 30%, and 50% of the demand, respectively. For each case, the resulting servicelevel and the average holding cost is given. Results in italics indicate that the target servicelevel of 95% is not met. In addition to the

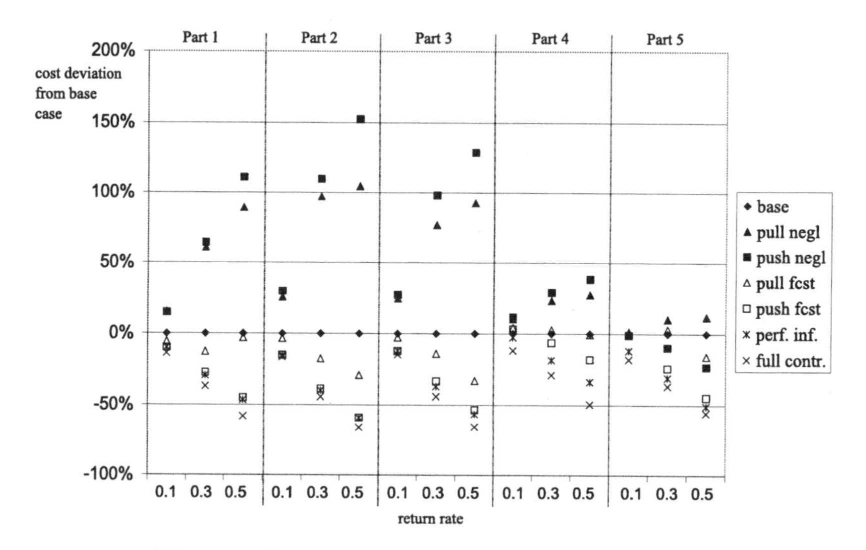

Fig. 9.4. Cost comparison for different return rates

Table 9.3. Simulation results for different return rates

		Part 1			Part 2			Part 3			Part 4			Part 5		
	dism. cover	**0.1**	**0.3**	**0.5**	**0.1**	**0.3**	**0.5**	**0.1**	**0.3**	**0.5**	**0.1**	**0.3**	**0.5**	**0.1**	**0.3**	**0.5**
base	**cost**	975	975	975	3000	3000	3000	3102	3102	3102	1092	1092	1092	4200	4200	4200
	servicel.	0.95	0.95	0.95	0.95	0.95	0.95	0.95	0.95	0.95	0.95	0.95	0.95	0.95	0.95	0.95
pull negl.	**cost**	1124	1568	1846	3764	5914	6135	3862	5487	5977	1204	1345	1389	4260	4625	*4693*
	servicel.	0.99	1.00	1.00	0.99	1.00	1.00	0.99	1.00	1.00	0.96	0.97	0.97	0.95	0.95	*0.94*
push negl.	**cost**	1119	1602	2056	3892	6290	7575	3945	6147	7100	1218	1408	1512	4164	3795	3223
	servicel.	0.99	1.00	1.00	0.99	1.00	1.00	0.99	1.00	1.00	0.97	0.99	0.99	0.96	0.98	0.98
pull net	**cost**	*906*	*783*	*789*	2892	*2487*	*1913*	*2932*	*2421*	*1899*	*1031*	*850*	*744*	4200	*3605*	*1476*
	servicel.	*0.94*	*0.93*	*0.87*	0.95	*0.94*	*0.94*	*0.94*	*0.93*	*0.93*	*0.94*	*0.90*	*0.89*	0.95	*0.94*	*0.87*
push net	**cost**	863	*642*	*467*	2546	*1761*	*1110*	*2645*	*1875*	*1207*	*972*	*814*	*699*	*3351*	*2616*	*1955*
	servicel.	0.95	*0.93*	*0.93*	0.95	*0.94*	*0.94*	*0.94*	*0.94*	*0.93*	*0.93*	*0.91*	*0.91*	*0.93*	*0.92*	*0.92*
pull fcst	**cost**	922	852	945	2892	2476	2126	3016	2667	2083	1139	1123	1088	4200	4332	3532
	servicel.	0.95	0.95	0.95	0.95	0.95	0.95	0.95	0.95	0.95	0.95	0.95	0.95	0.95	0.95	0.95
push fcst	**cost**	879	709	537	2546	1841	1233	2722	2070	1445	1128	1024	895	4164	3189	2324
	servicel.	0.95	0.95	0.95	0.95	0.95	0.95	0.95	0.95	0.95	0.95	0.95	0.95	0.95	0.95	0.95
perf. inf.	**cost**	878	690	522	2542	1804	1225	2727	1957	1349	1064	889	724	3701	2908	2072
	servicel.	0.95	0.95	0.95	0.95	0.95	0.95	0.95	0.95	0.95	0.95	0.95	0.95	0.95	0.95	0.95
full contr.	**cost**	840	616	410	2509	1674	1023	2645	1735	1069	963	778	552	3444	2661	1862
	servicel.	0.95	0.95	0.95	0.95	0.95	0.95	0.95	0.95	0.95	0.95	0.95	0.95	0.95	0.95	0.95

above policies, we include two scenarios assessing in more detail the value of information concerning product returns. As in Section 7.5 we consider the cases that product returns are known beforehand ('perfect information') and that timing of returns can be influenced ('full control'). In both cases we assume a dismantling push strategy. Figure 9.4 displays per part the cost deviation with respect to the base case for all policies meeting the servicelevel constraint.

We see essentially the same results as discussed above. Not surprisingly, the performance differences between the alternative policies increase with the average dismantling contribution. Moreover, they tend to be the more important the larger the product return volume during a traditional channel leadtime period. While cost deviations are within +/- 50% for the low demand parts 4 and 5, they increase beyond 100% for the high volume parts.

It is worth noting that the additional benefit of certainty or even control concerning product returns appears to be rather small here. In most cases costs of the 'push forecasting' policy exceed those under 'perfect information' by less than 10% and those under 'full control' by some 20%. One may expect this result to be a consequence of the fairly smooth return process and, in particular, of the small return batches. In the case of a more erratic return process involving large product batches the benefits of an active return policy may be more prominent.

9.4 Recommendations

Based on our findings we come to the following recommendations concerning the integration of dismantling into IBM's spare parts management. The above results emphasize the importance of an integral logistics process design. Merely adding dismantling as an additional parts supply option without adapting the control policy of the existing sources results in poor logistics performance where high stock levels partly cancel out the procurement cost benefits.

Explicitly taking into account future dismantling volumes turns out to be key to establishing an efficient logistics concept. Reliable forecasting information results in a substantial reduction of stocklevels even for moderate product return rates. Information is the more important the higher the potential dismantling volume during the leadtime period of the alternative sources. Given these results, taking a serious effort to build up a reliable base of information concerning the product return processes is strongly recommended. We have seen that probabilistic information on the return process may be sufficient from a logistics control perspective and that the additional benefit of detailed channel monitoring may be limited. However, it should be noted that estimating the relevant return process parameters may be facilitated significantly by a more active role in the return channel. This may be a major argument for adopting an active product return policy rather than perceiving the inflow of used products as an external given.

With respect to the design of the dismantling channel one may distinguish several situations. If parts dismantled from used equipment require largely the same processing as repairable exchange parts returned from the field and incur approximately the same costs then adding them to the existing AFR stock may be the easiest solution. In this case, incorporating dismantling is a matter of adapting information systems in the first place. Otherwise, dismantling should be established as a distinct channel. It turns out that the cost advantage of postponing the processing of dismantled parts until they are needed by means of an intermediate stockpoint is rather small and may often even be more than offset by an increase in the overall stock level. Taking into account the administrative cost for managing an additional stockpoint, a pull controlled dismantling channel therefore does not seem appropriate. Instead, dismantled parts should be pushed immediately into the CSP stock. It should be noted that this result also raises some questions concerning the appropriateness of the current AFR–stock concept. In principle, the above reasoning also applies to the repair channel. Reconsidering the corresponding logistics processes may therefore be worthwhile.

Finally, we conclude that inventory cost considerations hardly appear to be a valid reason for disposing of reusable parts. We have seen that procurement cost savings largely outweigh the additional holding costs for dismantled parts in many cases. Therefore, one may expect disposal only to become relevant at the end of a product lifecycle, when there is a risk of dismantled parts not to be reused at all. To this end, dismantling for a given part should be terminated once the available inventory is sufficient to cover the remainder of the service horizon. Otherwise, all parts for which a dismantling process is available should be recovered. This information can be easily implemented in the form of a 'keep list' analogous with the selection of machines for refurbishment (compare Chapter 2). The above results show that the major tradeoff determining whether or not to dismantle a certain part concerns the investment costs for setting up the necessary test processes, rather than additional holding costs.

10. Conclusions

We have begun our investigation in Chapter 1 by pointing out that goods flows in today's business environments are no longer unidirectional. Rather than following a clearly ordered hierarchy, supply chain members form general networks and the corresponding goods flows involve loops. Goods flows opposite to the traditional supply chain direction becoming increasingly important has given rise to the notion of 'Reverse Logistics'. In Chapter 3 we have indicated several categories of such 'reverse' flows. Although including traditional elements such as warranty returns and by–product recycling the recent interest in Reverse Logistics is mainly driven by the growing importance of used product and commercial returns. Throughout this monograph we have illustrated the relevance of reverse goods flows to companies' business activities by giving numerous examples stretching from copier remanufacturing to carpet recycling and from single–use cameras to reusable packaging. In particular, we have taken a closer look at return flow management at IBM, which we have seen to include used equipment recovery on the product, component, and material level. To round up our investigation we now return to the research questions formulated in Chapter 1.

Which logistics issues arise in the management of 'reverse' goods flows?
In Chapters 4 and 6 we have analysed numerous case studies illustrating current business practice. We have seen that 'reverse' goods flows are typically managed by the receiving party and hence that Reverse Logistics is a form of inbound logistics. Moreover, we have observed that the overall logistics task that companies face in this context is about bridging the gap between a former owner, releasing a product, and another future owner.

In Chapters 4 and 5 we have focussed on the spatial aspect of this gap, which led us to distribution management issues in Reverse Logistics. We have seen that many standard logistics tasks, such as transportation planning and vehicle routing, also arise in a Reverse Logistics context. In particular, logistics network design is an important issue. Companies need to take decisions on appropriate locations for collection and inspection sites and processing facilities for the returned products. Moreover, they strive for a good choice between consolidation and separation of the arising goods flows.

In Chapters 6–8 we have considered the temporal aspect of Reverse Logistics, namely inventory management issues. This concerns the timing of product recovery activities, in the first place. We have discussed that companies are looking for decisions on when to carry out the required inspection, disassembly, or repair steps of their reverse channel in order to strike a good tradeoff between processing costs and customer service. In particular, a balance is to be made between supply– and demand–driven activities. To this end, one needs to decide on the deployment of inventory to decouple individual transformation and transportation processes.

What are the differences between Reverse Logistics and traditional 'forward' logistics?
We conclude from our analysis that the distinction between 'forward' and 'reverse' logistics is not that much a matter of 'direction'. The differences between individual forward and reverse flows, between distribution and collection, between assembly and disassembly appear to be rather limited and do not lead to essentially different logistics decision problems. What *does* make Reverse Logistics different is the interaction between forward and reverse flows.

We have seen that a company's 'reverse' inbound flows in most cases entail subsequent 'forward' outbound flows. Therefore, logistics co–ordination must take into account market conditions both on the demand and the supply side. In conventional supply chain theory it is the end customer's demand that drives the entire chain. Supply is an endogenous variable which each party decides upon according to its needs. In a Reverse Logistics context, however, boundary conditions on the supply side are much more restrictive. At present, timing, quantity and quality of 'reverse' inbound flows are largely exogenously determined and may, in addition, be difficult to forecast. This observation is the most evident in the case of legislative take–back obligations. However, even purely economically motivated reverse flows form a significantly less homogenous input resource than conventional 'virgin' supply.

The above argumentation supports the frequently heard claim that Reverse Logistics is characterised by a high level of uncertainty. A lack of information on various aspects of reverse goods flows is, indeed, one of the difficulties that companies struggle with in their Reverse Logistics initiatives. However, we also see that uncertainty is only one aspect of a more general phenomenon. Even if product returns were completely known beforehand difficulties might arise from an imbalance between exogenous supply and demand forces. Therefore, we conclude more broadly that it is a lack of supply control, that complicates Reverse Logistics management.

To some extent the above characteristics surely reflect a general lack of experience in the young field of Reverse Logistics where many companies still follow a fairly reactive policy. As reverse goods flows are becoming more important one may expect companies to engage in a more active role towards return management rather than to perceive reverse flows as an external given. In particular, recent advances in electronic information technology may have a substantial impact in this context. For example, remote product monitoring and sensoring may offer opportunities for collecting extensive return data early in the process and hence reduce uncertainty significantly. This again may provide the means for a more advanced return management actively controlling the resource potential of 'reverse' goods flows. However, given that the goods in a 'reverse' flow are, by definition, derivatives of previous business activities rather than being limited to their role as future input resources one may expect some tension between supply and demand in Reverse Logistics to remain.

How can the characteristics of Reverse Logistics appropriately be captured in quantitative models that support decision making?
In several examples it has been shown how traditional logistics models can be adapted to a Reverse Logistics context. Following the above argumentation decisions should not be based on Reverse Logistics considered in isolation but should view the overall logistics context. As a consequence, the modelling scope may be broader than in some of the traditional domains. In this sense, Reverse Logistics models concur with recent supply chain management philosophy. The limited supply control in Reverse Logistics typically translates into additional restrictions in the corresponding quantitative models. To some extent, they can be interpreted as capacity restrictions. To capture the inherent uncertainty stochastic models may be a natural choice. However, we have seen that the impact of stochastic variations on solutions to the relevant logistics problems is very much dependent on context. Therefore, more simple deterministic approaches may sometimes be preferable.

In Chapter 5 a modelling framework has been introduced for logistics network design in a Reverse Logistics context. We have shown how traditional multi–level warehouse location models can be adapted accordingly. Moreover, we have seen that stochastic variations of the inbound 'reverse' flows often have rather limited effect on this strategic decision problem. Furthermore, we have explained under which conditions the overall network design problem may be decomposed into separate 'forward' and 'reverse' parts.

Chapters 7 and 8 have analysed inventory control models in a Reverse Logistics setting. Exogenous inbound flows have been shown to entail a loss of monotonicity in the mathematical models, which complicates their analysis. In addition, the selection from multiple supply alternatives has been identified as another major distinction with traditional models. However, it has been shown that these models can be transformed into equivalent standard models under some conditions. Furthermore, we have pointed out that on this operational level variations in the 'reverse' goods flow have, indeed, a major impact on logistics costs and need to be taken into account explicitly to avoid poor decision making.

We conclude that traditional operational research logistics models provide a good starting point, which can be adapted to develop appropriate Reverse Logistics models.

Having said this, it is time to look ahead again and to consider open issues appearing on the research agenda. Research on Reverse Logistics is surely still far from exhaustive. Additional case study material would be more than welcome to further refine the picture. Specifically, we would like to draw attention to the following issues.

- Probably the most challenging issue concerns the shift towards an *active return management*. As discussed above one may expect companies to look for ways to increase control over inbound goods flows as secondary resources gain further importance and Reverse Logistics develops into a standard supply chain element. Another reason for this development is the altering view on a company's products, which are increasingly perceived as a comprehensive set of services rather than being limited to physical goods. Reverse Logistics flows for repair and end–of–use handling are integral parts of such a product concept. Systematic research addressing opportunities to facilitate the development towards an active return management would be highly relevant. In particular, analysing the impact of modern information technology in this context appears to be more than worthwhile: which data should be collected and how can it be exploited in an optimal way?
- *Return forecasting* is a somewhat related issue. In most of our models we have assumed information on the stochastic properties of product returns to be known. In practice, however, the relevant parameters are not directly observable in many cases and need to be estimated instead. This issue has recently been addressed in detail in the context of single–use cameras (Toktay et al., 1999). Earlier investigations concern refillable bottles (Goh and Varaprasad, 1986; Kelle and Silver, 1989). However, as discussed in Chapter 6 these approaches are primarily applicable for short return cycles. For durable products which are only returned after a market-sojourn time of several years other methods appear to be needed to estimate relevant return data.
- Another relevant topic that has been largely neglected concerns returns *dispositioning*. Today, most Reverse Logistics models focus on a single product recovery option. However, as we have seen many companies deal with several recovery alternatives, such as product, component, and material reuse. Dispositioning is often based on a simple fixed hierarchy, e.g., preferring product reuse to component reuse. However, this approach may be far from optimal: Extracting an expensive component which can serve as a spare part may yield larger savings than remanufacturing a complete machine which eventually turns out to be non–remarketable. A systematic analysis of more advanced dispositioning strategies would therefore be desirable.

- Although including examples of different categories of 'reverse' goods flows our analysis is somewhat biased towards end–of–use product returns. To refine our findings an additional focus on *commercial returns* specifically would be valuable. Empirical evidence provided by Rogers and Tibben–Lembke (1999) appears to be a promising starting point in this direction. To the best of our knowledge corresponding quantitative analyses are few.
- Quantitative Reverse Logistics models currently available, as discussed in this monograph, largely focus on external logistics. In contrast, quantitative results on the impact of 'reverse' goods flows on *internal logistics* issues such as warehouse design and internal routing are largely lacking. Since empirical evidence indicates these issues to be a major concern in the practical implementation of Reverse Logistics initiatives (see, e.g., Stock, 1998; Rogers and Tibben–Lembke, 1999) a systematic analysis of the corresponding tradeoffs would surely be relevant.
- We have argued in the introduction that quantitative research in Reverse Logistics in this early stage should focus on developing and analysing appropriate models. Consequently, this is where we have put the emphasis of our contribution. As the field matures *computational aspects* may also deserve attention. As we have seen Reverse Logistics models tend to have a fairly large scope and may therefore be computationally expensive. Therefore, developing efficient solution techniques may be helpful to make Reverse Logistics models easy to apply.

Considering the above issues it is clear that Reverse Logistics offers many challenging opportunities both from a research and a business practice perspective. Hopefully, the research presented in this monograph can contribute to stimulating further developments.

List of Figures

List of Tables

References

1. Abramowitz, M. and Stegun, I.A., editor (1970). *Handbook of Mathematical Functions.* Dover, New York, N.Y.
2. Akinc, U. (1985). Multi–activity facility design and location problems. *Management Science*, 31(3):275–283.
3. Ammons, J.C., Realff, M.J., and Newton, D.J. (1997). Reverse production system design and operation for carpet recycling. Working paper, Georgia Institute of Technology.
4. Anonymous (1998). Product return flows & physical distribution complaints: Critical success factors for organizing both processes. Master's thesis, Erasmus University Rotterdam, The Netherlands.
5. APICS (1998). Remanufacturing Seminar 1997/98 combined year proceedings.
6. Arola, D.F., Allen, L.E., and Biddle, M.B. (1999). Evaluation of mechanical recycling options for electronic equipment. In *Proceedings of the IEEE International Symposium on Electronics and the Environment*, pages 187–191, Danvers, MA.
7. Arrow, K.J., Karlin, S., and Scarf, H. (1958). *Studies in the Mathematical Theory of Inventory and Production.* Stanford University Press, Stanford, CA.
8. Ayres, R.U., Ferrer, G., and Van Leynseele, T. (1997). Eco–efficiency, asset recovery and remanufacturing. *European Management Journal*, 15(5):557–574.
9. Barros, A.I., Dekker, R., and Scholten, V. (1998). A two–level network for recycling sand: A case study. *European Journal of Operational Research*, 110:199–214.
10. Beltrán, J.L., Beyer, D., Krass, D., and Sridhar, R. (1997). Stochastic inventory models with returns and delivery commitments. Working paper, University of Toronto, Faculty of Management, Toronto, Canada.
11. Beltrán, J.L. and Krass, D. (1997). Dynamic lot sizing with returning items and disposals. Working paper, University of Toronto, Faculty of Management, Toronto, Canada. (To appear in IIE Transactions).
12. Berger, T. and Debaillie, B. (1997). Location of disassembly centres for re–use to extend an existing distribution network. Master's thesis, University of Leuven, Belgium. (In Dutch).
13. Beullens, P., Van Oudheusden, D., and Cattrysse, D. (1999a). Bi–destination waste collection: Impact of vehicle type and operations on transportation costs. In Flapper, S.D.P. and De Ron, A.J., editors, *Proceedings of the Second International Working Seminar on Reuse*, pages 5–14, Eindhoven, The Netherlands.
14. Beullens, P., Van Wassenhove, L.N., Van Oudheusden, D., and Cattrysse, D. (1999b). An analysis of the combined routing of the collection

of used products and the distribution of new products. In Van Goor, A.R., Flapper, S.D.P., and Clement, C., editors, *Reverse Logistics*, chapter B5200. Kluwer, The Netherlands. (In Dutch).
15. BLOEMHOF-RUWAARD, J.M. (1996). *Integration of Operations Research and Environmental Management.* PhD thesis, University of Wageningen, The Netherlands.
16. BLOEMHOF-RUWAARD, J.M., FLEISCHMANN, M., AND VAN NUNEN, J.A.E.E. (1999). Reviewing distribution issues in reverse logistics. In Speranza, M.G. and Stähly, P., editors, *New Trends in Distribution Logistics*, pages 23–44. Springer, Berling, Germany.
17. BLOEMHOF-RUWAARD, J.M. AND SALOMON, M. (1997). Reverse Logistics. In Ploos van Amstel, M.J., Duijker, J.P., and de Koster, M.B.M., editors, *Praktijkboek Magazijnen en Distributiecentra.* Kluwer Bedrijfswetenschappen, Deventer, The Netherlands. (In Dutch).
18. BLOEMHOF-RUWAARD, J.M., VAN WASSENHOVE, L.N., GABEL, H.L., AND WEAVER, P.M. (1996). An environmental life cycle optimization model for the european pulp and paper industry. *Omega*, 24(6):615–629.
19. BRAS, B. AND MCINTOSH, M.W. (1999). Product, process, and organizational design for remanufacture: an overview of research. *Robotics and Computer Integrated Manufacturing*, 15:167–178.
20. BRENNAN, L., GUPTA, S.M., AND TALEB, K.N. (1994). Operations planning issues in an assembly/disassembly environment. *International Journal of Operations and Production Management*, 14(9):57–67.
21. BROWN-HUMES, C. (1999). Brittle balance of recycling. *Financial Times*, Dec 8.
22. BUCHANAN, D.J. AND ABAD, P.L. (1998). Optimal policy for a periodic review returnable inventory system. *IIE Transactions*, 30:1049–1055.
23. CAIRNCROSS, F. (1992). How Europe's companies reposition to recycle. *Harvard Business Review*, 70(2):34–45.
24. CANON (1998). Annual corporate environmental report.
25. CARTER, C.R. AND ELLRAM, L.M. (1998). Reverse logistics: A review of the literature and framework for future investigation. *International Journal of Business Logistics*, 19(1):85–102.
26. CBS (1999). Statistical yearbook for The Netherlands.
27. CHANDRASHEKAR, A. AND DOUGLESS, T.C. (1996). Commodity indexed surplus asset disposal in the reverse logistics process. *The International Journal of Logistics Management*, 7(2):59–68.
28. CHEN, S.X. AND LAMBRECHT, M. (1996). X–Y band and modified (s, S) policy. *Operations Research*, 44(6):1013–1019.
29. CHO, D.I. AND PARLAR, M. (1991). A survey of maintenance models for multi-unit systems. *European Journal of Operational Research*, 51:1–23.
30. CLEGG, A., WILLIAMS, D., AND UZSOY, R. (1995). Production planning for companies with remanufacturing capability. In *Proceedings of the IEEE Symposium on Electronics and the Environment*, pages 186–191, Orlando, Florida.
31. CLENDENIN, J.A. (1997). Closing the supply chain loop: Reengineering the returns channel process. *International Journal of Logistics Management*, 8(1):75–85.
32. COHEN, M.A., NAHMIAS, S., AND PIERSKALLA, W.P. (1980). A dynamic inventory system with recycling. *Naval Research Logistics Quarterly*, 27(2):289–296.
33. CONSTANTINIDES, G.M. (1976). Stochastic cash management with fixed and proportional transaction costs. *Management Science*, 22(12):1320–1331.

34. Corbey, M., Inderfurth, K., van der Laan, E.A., and Minner, S. (1999). The use of accounting information in production and inventory control for reverse logistics. Working Paper 24/99, University of Magdeburg, Germany.
35. Crainic, T.G., Gendreau, M., and Dejax, P. (1993). Dynamic and stochastic models for the allocation of empty containers. *Operations Research*, 41(1):102–126.
36. Crainic, T.G. and Laporte, G. (1997). Planning models for freight transportation. *European Journal of Operational Research*, 97:409–438.
37. Daganzo, C.F. (1999). *Logistics Systems Analysis*. Springer, Berlin, 3rd edition.
38. Daskin, M.S. (1995). *Network and Discrete Location*. Wiley, Chichester.
39. de Koster, M.B.M., Krikke, H.R., Flapper, S.D.P., and Vermeulen, W.S. (1999). Reverse logistics in the white goods sector. Working paper, Erasmus University Rotterdam, The Netherlands. (In Dutch).
40. de Koster, M.B.M. and van de Vendel, M. (1999). Efficient return handling in the retail: A comparison. Working paper, Erasmus University Rotterdam, The Netherlands.
41. Dijkhuizen, H.P. (1997). Reverse Logistics bij IBM. In Van Goor, A.R., Flapper, S.D.P., and Clement, C., editors, *Reverse Logistics*, chapter E1310. Kluwer, The Netherlands. (In Dutch).
42. Dillon, P.S. (1994). 1994. *IEEE Spectrum*, 31(8):18–21.
43. Driesch, H.M., van Oyen, J.E., and Flapper, S.D.P. (1997). Control of the Daimler–Benz MTR product recovery operation. Working paper, Eindhoven University of Technology, The Netherlands.
44. DSM (1999). Press releases. Apr 26 and Nov 15.
45. Duales System Deutschland (2000). http://www.gruener-punkt.de/en/. (May 31, 2000).
46. Dupont (1999). Annual corporate environmental report.
47. Eiselt, H.A., Gendreau, M., and Laporte, G. (1995). Arc routing problems, part2: The rural postman problem. *Operations Research*, 43:399–414.
48. Elmendorp, P. (1998). Controlling or controlled by reusable packaging. Master's thesis, Erasmus University Rotterdam, The Netherlands. (In Dutch).
49. Emmons, H. and Gilbert, S.M. (1998). The role of returns policies in pricing and inventory decisions for catalogue goods. *Management Science*, 44(2):276–283.
50. Erlenkotter, D. (1978). A dual–based procedure for uncapacitated facility location. *Operations Research*, 26:992–1009.
51. Eskigun, E. and Uzsoy, R. (1998). Design and control of supply chains with product recovery and remanufacturing. Working paper, Purdue University, West Lafayette, IN.
52. Federgruen, A. and Zheng, Y.S. (1992). An efficient algorithm for computing an optimal (r,Q) policy in continuous review stochastic inventory systems. *Operations Research*, 40:808–813.
53. Ferrer, G. (1996). Product recovery management: Industry practices and research issues. Working Paper 96/55/TM, INSEAD, Fontainebleau, France.
54. Ferrer, G. (1997a). The economics of personal computer remanufacturing. *Resources, Conservation and Recycling*, 21(2):79–108.
55. Ferrer, G. (1997b). The economics of tire remanufacturing. *Resources, Conservation and Recycling*, 19(4):221–225.
56. Flapper, S.D.P. (1994b). Matching material requirements and availabilities in the context of recycling: An MRP–I based heuristic. In *Proceedings of*

the Eighth International Working Seminar on Production Economics, pages 511–519 (Vol.3), Igls/Innsbruck, Austria.

57. FLAPPER, S.D.P. (1996a). Logistic aspects of reuse. In Flapper, S.D.P. and de Ron, A.J., editors, *Proceedings of the First International Working Seminar on Reuse*, pages 109–118, The Netherlands. Eindhoven University of Technology.
58. FLAPPER, S.D.P. (1996b). One–way or reusable distribution items? In *Proceedings of the Second International Conference on Computer Integrated Manufacturing in the Process Industries*, pages 230–243, Eindhoven. Institute for Business Engineering and Technology Application.
59. FLAPPER, S.D.P. AND JENSEN, T. (1998). Logistic planning and control of rework. Research report, University of Eindhoven, The Netherlands.
60. FLAPPER, S.D.P. AND DE RON, A.J., editor (1996). *Proceedings of the First International Working Seminar on Reuse*, The Netherlands. Eindhoven University of Technology.
61. FLAPPER, S.D.P. AND DE RON, A.J., editor (1999). *Proceedings of the Second International Working Seminar on Reuse*, The Netherlands. Eindhoven University of Technology.
62. FLEISCHMANN, M., BEULLENS, P., BLOEMHOF-RUWAARD, J.M., AND VAN WASSENHOVE, L.N. (2000). The impact of product recovery on logistics network design. Working paper, INSEAD, Fontainebleau, France.
63. FLEISCHMANN, M., BLOEMHOF-RUWAARD, J.M., DEKKER, R., VAN DER LAAN, E.A., VAN NUNEN, J.A.E.E., AND VAN WASSENHOVE, L.N. (1997a). Quantitative models for reverse logistics: A review. *European Journal of Operational Research*, 103:1–17.
64. FLEISCHMANN, M., KRIKKE, H.R., DEKKER, R., AND FLAPPER, S.D.P. (1999). A characterisation of logistics networks for product recovery. Management Report Series 17(99), Erasmus University Rotterdam, The Netherlands. (To appear in Omega).
65. FLEISCHMANN, M. AND KUIK, R. (1998). On optimal inventory control with stochastic item returns. Management Report Series 21-98, Erasmus University Rotterdam, The Netherlands.
66. FLEISCHMANN, M., KUIK, R., AND DEKKER, R. (1997b). Controlling inventories with stochastic item returns: A basic model. Management Report Series 43(13), Erasmus University Rotterdam, The Netherlands. (To appear in the European Journal of Operational Research).
67. FULLER, D.A. AND ALLEN, J. (1995). A typology of reverse channel systems for post-consumer recyclables. In Polonsky, J. and Mintu-Winsatt, A.T., editors, *Environmental Marketing: Strategies, Practice, Theory, and Research*, pages 241–266. Haword Press, Binghamton, NY.
68. GABEL, H.L., WEAVER, P.M., BLOEMHOF-RUWAARD, J.M., AND VAN WASSENHOVE, L.N. (1996). Life cycle analysis and policy options: The case of the european pulp and paper industry. *Business Strategy and the Environment*, 5:156–167.
69. GANESHAN, R., JACK, E., MAGAZINE, M.J., AND STEPHENS, R. (1998). A taxonomic review of supply chain management research. In Tayur, S.R., Ganeshan, R., and Magazine, M.J., editors, *Quantitative Models for Supply Chain Management*, chapter 27. Kluwer.
70. GEOFFRION, A.M. AND NAUSS, R. (1977). Parametric and postoptimality analysis in integer linear programming. *Management Science*, 23:453–466.
71. GINTER, P.M. AND STARLING, J.M. (1978). Reverse distribution channels for recycling. *California Management Review*, 20(3):73–82.

72. GLASSEY, R. AND GUPTA, V. (1975). An LP analysis of paper recycling. In Salkin, H. and Saha, J., editors, *Studies in Linear Programming*, pages 273–292. North–Holland.
73. GOH, T.N. AND VARAPRASAD, N. (1986). A statistical methodology for the analysis of the life–cycle of reusable containers. *IIE Transactions*, 18:42–47.
74. GOTZEL, C., WEIDLING, J., HEISIG, G., AND INDERFURTH, K. (1999). Product return and recovery concepts of companies in Germany. Working Paper 31/99, University of Magdeburg, Germany.
75. GROENEWEGEN, P. AND DEN HOND, F. (1993). Product waste in the automotive industry: Technology and environmental management. *Business Strategy and the Environment*, 2(1):1–12.
76. GUIDE, JR., V.D.R. (1996). Scheduling using drum–buffer–rope in a remanufacturing environment. *International Journal of Production Research*, 34(4):1081–1091.
77. GUIDE, JR., V. (2000). Production planning and control for remanufacturing: Industry practice and research needs. *Journal of Operations Management*, 18:467–483.
78. GUIDE, JR., V.D.R., JAYARAMAN, V., SRIVASTAVA, R., AND BENTON, W.C. (1998). Supply chain management for remanufacturable manufacturing systems. Working Paper 98/101/OM, Suffolk University, Boston, MA. (To appear in Interfaces).
79. GUIDE, JR., V.D.R. AND SRIVASTAVA, R. (1997a). An evaluation of order release strategies in a remanufacturing environment. *Computers and Operations Research*, 24(1):37–48.
80. GUIDE, JR., V.D.R. AND SRIVASTAVA, R. (1997b). Repairable inventory theory: Models and applications. *European Journal of Operational Research*, 102:1–20.
81. GUIDE, JR., V.D.R. AND SRIVASTAVA, R. (1999). The effect of lead time variability on the performance of disassembly release mechanisms. *Computers and Industrial Engineering*, 36:759–779.
82. GUIDE, JR., V.D.R., SRIVASTAVA, R., AND KRAUS, M.E. (1997a). Product structure complexity and scheduling of operations in recoverable manufacturing. *International Journal of Production Research*, 35(11):3179–3199.
83. GUIDE, JR., V.D.R., SRIVASTAVA, R., AND SPENCER, M.S. (1997b). An evaluation of capacity planning techniques in a remanufacturing environment. *International Journal of Production Research*, 35(1):67–82.
84. GUILTINAN, J.P. AND NWOKOYE, N.G. (1975). Developing distribution channels and systems in the emerging recycling industries. *International Journal of Physical Distribution*, 6(1):28–38.
85. GUNGOR, A. AND GUPTA, S.M. (1999). Issues in environmentally conscious manufacturing and product recovery: A survey. *Computers and Industrial Engineering*, 36:811–853.
86. GUPTA, S.M. AND TALEB, K.N. (1994). Scheduling disassembly. *International Journal of Production Research*, 32(8):1857–1866.
87. HADLEY, G. AND WHITIN, T.M. (1963). *Analysis of Inventory Systems*. Prentice–Hall, Englewood Cliffs, N.J.
88. HAMMERSCHMID, R. (1990). *Development of Technically-Economically Optimised Regional Disposal Alternatives*. Physica–Verlag, Heidelberg, Germany. (In German).
89. HEISIG, G. AND FLEISCHMANN, M. (1999). Planning stability in a product recovery system. Working paper, University of Magdeburg, Germany. (To appear in OR Spektrum).

90. HEYMAN, D.P. (1977). Optimal disposal policies for a single–item inventory system with returns. *Naval Research Logistics Quarterly*, 24:385–405.
91. HEYMAN, D.P. AND SOBEL, M.J. (1984). *Stochastic Models in Operations Research*, volume II. McGraw–Hill, New York.
92. IBM (1998). Annual corporate environmental report.
93. IEEE (1999). Proceedings of the IEEE International Symposium on Electronics and the Environment.
94. IGLEHART, D. (1963). Optimality of (s, S) policies in the infinite–horizon dynamic inventory problem. *Management Science*, 9:259–267.
95. INDERFURTH, K. (1982). Zum Stand der betriebswirtschaftlichen Kassenhaltungstheorie. *Zeitschrift für Betriebswirtschaft*, 3:295–320. (In German).
96. INDERFURTH, K. (1996). Modeling period review control for a stochastic product recovery problem with remanufacturing and procurement leadtimes. Working paper 2/96, Faculty of Economics and Management, University of Magdeburg, Germany.
97. INDERFURTH, K. (1997). Simple optimal replenishment and disposal policies for a product recovery system with leadtimes. *OR Spektrum*, 19:111–122.
98. INDERFURTH, K. (1998). The performance of simple MRP driven policies for stochastic manufacturing/remanufacturing problems. Working Paper 32/98, University of Magdeburg, Germany.
99. INDERFURTH, K., DE KOK, A.G., AND FLAPPER, S.D.P. (1998). Product recovery policies in stochastic remanufacturing systems with multiple reuse options. Working Paper 13/98, University of Magdeburg, Germany.
100. INDERFURTH, K. AND JENSEN, T. (1998). Analysis of MRP policies with recovery options. Working paper, University of Magdeburg, Germany.
101. INDERFURTH, K. AND VAN DER LAAN, E.A. (1998). Leadtime effects and policy improvement for stochastic inventory control with remanufacturing. Working Paper 22/98, University of Magdeburg, Germany. (To appear in International Journal of Production Economics).
102. JAGDEV, S. (1999). The impact of reverse logistics on vehicle routing algorithms. Presentation at INFORMS Conference, Cincinnati.
103. JAHRE, M. (1995). *Logistics Systems for Recycling — Efficient Collection of Household Waste*. PhD thesis, Chalmers University of Technology, Göteborg, Sweden.
104. JAYARAMAN, V., GUIDE, JR., V.D.R., AND SRIVASTAVA, R. (1999). A closed–loop logistics model for remanufacturing. *Journal of the Operational Research Society*, 50(5):497–508.
105. JENKINS, L. (1982). Parametric mixed integer programming: An application to solid waste management. *Management Science*, 28(11):1270–1284.
106. JOHNSON, M.R. AND WANG, M.H. (1995). Planning product disassembly for material recovery opportunities. *International Journal of Production Research*, 33(11):3119–3142.
107. JOHNSON, M.R. AND WANG, M.H. (1998). Economical evaluation of disassembly operations for recycling, remanufacturing, and reuse. *International Journal of Production Research*, 36(12):3227–3252.
108. JOHNSON, P.F. (1998). Managing value in reverse logistics systems. *Transportation Research – E (Logistics and Transportation Review)*, 34(3):217–227.
109. KELLE, P. AND SILVER, E.A. (1989a). Forecasting the returns of reusable containers. *Journal of Operations Management*, 8(1):17–35.
110. KELLE, P. AND SILVER, E.A. (1989b). Purchasing policy of new containers considering the random returns of previously issued containers. *IIE Transactions*, 21(4):349–354.

111. KLAUSNER, M., GRIMM, W.M., AND HORVATH, A. (1999). Integrating product take–back and technical service. In *Proceedings of the IEEE International Symposium on Electronics and the Environment*, pages 48–53, Danvers, MA.
112. KODAK (1999). Annual corporate environmental report.
113. KOKKINAKI, A.I., DEKKER, R., VAN NUNEN, J.A.E.E., AND PAPPIS, C. (1999). An exploratory study on electronic commerce for reverse logistics. Working Paper 9951/A, Erasmus University Rotterdam, The Netherlands.
114. KOPICKY, R.J., BERG, M.J., LEGG, L., DASAPPA, V., AND MAGGIONI, C. (1993). *Reuse and Recycling: Reverse Logistics Opportunities.* Council of Logistics Management, Oak Brook, IL.
115. KRIKKE, H.R. (1998). *Recovery Strategies and Reverse Logistics Network Design.* PhD thesis, University of Twente, Enschede, The Netherlands.
116. KRIKKE, H.R., VAN HARTEN, A., AND SCHUUR, P.C. (1998). On a medium term product recovery and disposal strategy for durable assembly products. *International Journal of Production Research*, 36(1):111–139.
117. KRIKKE, H.R., VAN HARTEN, A., AND SCHUUR, P.C. (1999). Business case océ: Reverse logistic network re-design for copiers. *OR Spektrum*, 21(3):381–409.
118. KROON, L. AND VRIJENS, G. (1995). Returnable containers: An example of reverse logistics. *International Journal of Physical Distribution & Logistics Management*, 25(2):56–68.
119. KRUPP, J.A.G. (1993). Structuring bills of material for automotive remanufacturing. *Production and Inventory Management Journal*, 34(4):46–52.
120. LEE, C.-H. (1997). Management of scrap car recycling. *Resources, Conservation and Recycling*, 20:207–217.
121. LEE, C.-H., CHANG, C.-T., AND TSAI, S.-L. (1998). Development and implementation of producer responsibility recycling systems. *Resources, Conservation and Recycling*, 24:121–135.
122. LOUWERS, D., KIP, B.J., PEETERS, E., SOUREN, F., AND FLAPPER, S.D.P. (1999). A facility location allocation model for re-using carpet materials. *Computers and Industrial Engineering*, 36(4):855–869.
123. LUND, R. (1984). Remanufacturing. *Technology Review*, 87(2):19–29.
124. MABINI, M.C., PINTELON, L.M., AND GELDERS, L.F. (1992). EOQ type formulations for controlling repairable inventories. *International Journal of Production Economics*, 28:21–33.
125. MARÍN, A. AND PELEGRÍN, B. (1998). The return plant location problem: Modelling and resolution. *European Journal of Operational Research*, 104:375–392.
126. MEACHAM, A., UZSOY, R., AND VENKATADRI, U. (1999). Optimal disassembly configurations for single and multiple products. *Journal of Manufacturing Systems*, 18(5):311–322.
127. MEYN, S.P. AND TWEEDIE, R.L. (1993). *Markov Chains and Stochastic Stability.* Springer, London.
128. MINNER, S. AND KLEBER, R. (1999). Optimal control of production and remanufacturing. Working Paper 32/99, University of Magdeburg, Germany.
129. MIRCHANDANI, P.B. AND FRANCIS, R.L. (1989). *Discrete Location Theory.* Wiley Publications, New York.
130. MOINZADEH, K. AND NAHMIAS, S. (1988). A continuous review model for an inventory system with two supply modes. *Management Science*, 34(6):761–773.
131. MOYER, L.K. AND GUPTA, S.M. (1997). Environmental concerns and recycling/disassembly efforts in the electronics industry. *Journal of Electronics Manufacturing*, 7(1):1–22.

132. Muckstadt, J.A. and Isaac, M.H. (1981). An analysis of single item inventory systems with returns. *Naval Research Logistics Quarterly*, 28:237–254.
133. Nahmias, S. (1981). Managing repairable item inventory systems: A review. *TIMS Studies in the Management Sciences*, 16:253–277.
134. Nasr, N., Hughson, C., Varel, E., and Bauer, R. (1998). State–of–the–art assessment of remanufacturing technology. Draft document, Rochester Institute of Technology.
135. Newton, D.J., Realff, M.J., and Ammons, J.C. (1999). Carpet recycling: The value of cooperation and a robust approach to determining the reverse production system design. In Flapper, S.D.P. and de Ron, A.J., editors, *Proceedings of the Second International Working Seminar on Reuse*, pages 207–216, Eindhoven, The Netherlands.
136. Océ (1998). Annual corporate environmental report.
137. OECD (1999). OECD environmental data: Compendium 1999.
138. Panisset, B.D. (1988). MRP II for repair/refurbish industries. *Production and Inventory Management Journal*, 29(4):12–15.
139. Penev, K.D. and de Ron, A.J. (1996). Determination of a disassembly strategy. *International Journal of Production Research*, 34(2):495–506.
140. Pierskalla, W.P. and Voelker, J.A. (1976). A survey of maintainance models: the control and surveillance of deteriorating systems. *Naval Research Logistics Quarterly*, 23:353–388.
141. Pnueli, Y. and Zussman, E. (1997). Evaluating the end–of–life value of a product and improving it by redesign. *International Journal of Production Research*, 35(4):921–942.
142. Pohlen, T.L. and Farris II, M. (1992). Reverse logistics in plastic recycling. *International Journal of Physical Distribution & Logistics Management*, 22(7):35–47.
143. Püchert, H., Spengler, T., and Rentz, O. (1996). Strategic recycling and redistribution management - a case study for the scrap car recycling. *Zeitschrift für Planung*, 7:27–44. (In German).
144. Realff, M.J., Ammons, J.C., and Newton, D.J. (1999). Carpet recycling: Determining the reverse production system design. *The Journal of Polymer–Plastics Technology and Engineering*, 38:547–567.
145. Richter, K. (1994). An EOQ repair and waste disposal model. In *Proceedings of the Eighth International Working Seminar on Production Economics*, pages 83–91, Igls/Innsbruck, Austria.
146. Richter, K. (1996a). The EOQ repair and waste disposal model with variable setup numbers. *European Journal of Operational Research*, 95:313–324.
147. Richter, K. (1997). Pure and mixed strategies for the EOQ repair and waste disposal problem. *OR Spektrum*, 19:123–129.
148. Richter, K. and Sombrutzki, M. (2000). Remanufacturing planning for the reverse Wagner/Whitin models. *European Journal of Operational Research*, 121:304–315.
149. Rogers, D.S. and Tibben-Lembke, R.S. (1999). *Going Backwards: Reverse Logistics Trends and Practices.* Reverse Logistics Executive Council, Pittsburgh, PA.
150. Romijn, P.M. (1999). Return flows of electrical and electronic equipment in the netherlands. Master's thesis, Erasmus University Rotterdam, The Netherlands. (In Dutch).
151. Rosenau, W.V., Twede, D., Mazzeo, M.A., and Singh, S.P. (1996). Returnable/reusable logistical packaging: A capital budgeting investment decision framework. *International Journal of Business Logistics*, 17(2):139–165.

152. RUDI, N. AND PYKE, D.F. (1999). Product recovery at the Norwegian national insurance administration. *Interfaces.* (to appear).
153. SAVASKAN, R.C., BHATTACHARYA, S., AND VAN WASSENHOVE, L.N. (1999). Channel choice and coordination in a remanufacturing environment. Working Paper 99/14/TM, INSEAD, Fontainebleau, France.
154. SAVASKAN, R.C. AND VAN WASSENHOVE, L.N. (1999). Implications of product take–back on channel profits: The case of multiple retail outlets. Working paper, INSEAD, Fontainebleau, France.
155. SCARF, H. (1960). The optimality of (s, S) policies in the dynamic inventory problem. In Arrow, K., Karlin, S., and Suppes, P., editors, *Mathematical Models in Social Sciences.* Stanford University Press.
156. SCELSI, P. (1991). Cleaning the earth through logistics. *Distribution,* 90(12):56–58.
157. SCHRADY, D.A. (1967). A deterministic inventory model for repairable items. *Naval Research Logistics Quarterly,* 14:391–398.
158. SCHUT, R. AND GERMANS, R. (1997). Reverse Logistics bij IBM. In Ploos van Amstel, M.J., Duijker, J.P., and de Koster, M.B.M., editors, *Praktijkboek Magazijnen en Distributiecentra.* Kluwer Bedrijfswetenschappen, Deventer, The Netherlands. (In Dutch).
159. SENNOTT, L.I. (1989). Average cost optimal stationary policies in infinite state markov decision processes with unbounded costs. *Operations Research,* 37(4):626–633.
160. SHERBROOKE, C.C. (1968). Metric: A multi–echelon technique for recoverable item control. *Operations Research,* 16:122–141.
161. SILVER, E.A., PYKE, D.F., AND PETERSON, R. (1998). *Inventory Management and Production Planning and Scheduling.* John Wiley & Sons, New York, 3rd edition.
162. SIMPSON, V.P. (1970). An ordering model for recoverable stock items. *AIIE Transactions,* 2(4):315–320.
163. SIMPSON, V.P. (1978). Optimum solution structure for a repairable inventory problem. *Operations Research,* 26(2):270–281.
164. SODHI, M.S., YOUNG, J., AND KNIGHT, W.A. (1998). Modeling material separation processes in bulk recycling. *International Journal of Production Research.* (To appear).
165. SPENGLER, T., PÜCHERT, H., PENKUHN, T., AND RENTZ, O. (1997). Environmental integrated production and recycling management. *European Journal of Operational Research,* 97:308–326.
166. SPENGLER, T. AND RENTZ, O. (1996). Planning models for the dismantling and recycling of complex mixed products. *Zeitschrift für Betriebswirtschaft,* 96(2):79–96. (In German).
167. STOCK, J.R. (1992). *Reverse Logistics.* Council of Logistics Management, Oak Brook, IL.
168. STOCK, J.R. (1998). *Development and Implementation of Reverse Logistics Programs.* Council of Logistics Management, Oak Brook, IL.
169. STUART, J.A., AMMONS, J.C., AND TURBINI, L.J. (1999). A product and process selection model with multidisciplinary environmental considerations. *Operations Research,* 47(2):221–234.
170. TALEB, K. AND GUPTA, S. (1997). Disassembly of multiple product structures. *Computers and Industrial Engineering,* 32(4):949–961.
171. TAYUR, S.R. AND GANESHAN, R. AND MAGAZINE, M.J., editor (1998). *Quantitative Models for Supply Chain Management.* Kluwer.

172. TEUNTER, R.H. (1998). Economic ordering quantities for remanufacturable item inventory systems. Working Paper 31/98, University of Magdeburg, Germany.
173. TEUNTER, R.H. (1999). Economic ordering quantities for stochastic inventory systems with reverse logistics. Working Paper 11/99, University of Magdeburg, Germany.
174. TEUNTER, R.H. AND INDERFURTH, K. (1998). The 'right' holding cost rates in average cost inventory models with reverse logistics. Working Paper 28/98, University of Magdeburg, Germany.
175. TEUNTER, R., VAN DER LAAN, E., AND INDERFURTH, K. (2000). How to set the holding cost rates in average cost inventory models with reverse logistics. *Omega*, 28(4):409–415.
176. THIERRY, M.C. (1997). *An Analysis of the Impact of Product Recovery Management on Manufacturing Companies.* PhD thesis, Erasmus University Rotterdam, The Netherlands.
177. THIERRY, M.C., SALOMON, M., VAN NUNEN, J.A.E.E., AND VAN WASSENHOVE, L.N. (1995). Strategic issues in product recovery management. *California Management Review*, 37(2):114–135.
178. TIBBEN-LEMBKE, R.S. (1999). The impact of reverse logistics on the total cost of ownership. *Journal of Marketing: Theory and Practice*, pages 51–60.
179. TIJMS, H. (1994). *Stochastic Models. An Algorithmic Approach.* Wiley, Chichester, UK.
180. TOKTAY, L.B., WEIN, L.M., AND ZENIOS, S.A. (1999). Inventory management of remanufacturable products. Working paper, INSEAD. (To appear in Management Science).
181. TOTH, P. AND VIGO, D. (1999). A heuristic algorithm for the symmetric and assymmetric vehicle routing problems with backhauls. *European Journal of Operational Research*, 113:528–543.
182. TRUNK, C. (1993). Making ends meet with returnable plastic containers. *Material Handling Engineering*, 48(10):79–85.
183. TSAY, A.A., NAHMIAS, S., AND AGRAWAL, N. (1998). Modeling supply chain contracts: A review. In Tayur, S., Ganeshan, R., and Magazine, M., editors, *Quantitative Models in Supply Chain Management*, chapter 27. Kluwer.
184. VAN DER LAAN, E.A. (1997). *The Effects of Remanufacturing on Inventory Control.* PhD thesis, Erasmus University Rotterdam, The Netherlands.
185. VAN DER LAAN, E.A. (1999). An NPV and AC analysis of a stochastic inventory system with joint manufacturing and remanufacturing. Working paper, Erasmus University Rotterdam, The Netherlands.
186. VAN DER LAAN, E.A., DEKKER, R., AND SALOMON, M. (1996a). An (s, Q) inventory model with remanufacturing and disposal. *International Journal of Production Economics*, 46–47:339–350.
187. VAN DER LAAN, E.A., DEKKER, R., AND SALOMON, M. (1996b). Product remanufacturing and disposal: A numerical comparison of alternative control strategies. *International Journal of Production Economics*, 45:489–498.
188. VAN DER LAAN, E.A., FLEISCHMANN, M., DEKKER, R., AND VAN WASSENHOVE, L.N. (1998). Inventory control for joint manufacturing and remanufacturing. In Tayur, S.R., Ganeshan, R., and Magazine, M.J., editors, *Quantitative Models in Supply Chain Management*, chapter 26. Kluwer.
189. VAN DER LAAN, E.A. AND SALOMON, M. (1997). Production planning and inventory control with remanufacturing and disposal. *European Journal of Operational Research*, 102:264–278.

190. VAN DER LAAN, E.A., SALOMON, M., AND DEKKER, R. (1999a). Leadtime effects in push and pull controlled manufacturing/remanufacturing systems. *European Journal of Operational Research*, 115(1):195–214.
191. VAN DER LAAN, E.A., SALOMON, M., DEKKER, R., AND VAN WASSENHOVE, L.N. (1999b). Inventory control in hybrid systems with remanufacturing. *Management Science*, 45(5):733–747.
192. VANDERMERWE, S. AND OLIFF, M.D. (1991). Corporate challenges for an age of reconsumption. *The Columbia Journal of World Business*, 26(3):7–25.
193. VAN GOOR, A.R. AND FLAPPER, S.D.P. AND CLEMENT, C., editor (1997). *Reverse Logistics*. Kluwer, Deventer, The Netherlands.
194. VEINOTT, JR., A.F. (1966). On the optimality of (s, S) inventory policies: New conditions and a new proof. *SIAM Journal of Applied Mathematics*, 14(5):1067–1083.
195. VROM (2000). Dutch ministry of housing, spatial planning and the environment. http://www.minvrom.nl/environment/. (May 31, 2000).
196. WAGNER, H.M. AND WHITIN, T.M. (1958). Dynamic version of the economic lot size model. *Management Science*, 5:212–219.
197. WANG, C.-H., EVEN, JR., J.C., AND ADAMS, S.K. (1995). A mixed-integer linear model for optimal processing and transport of secondary materials. *Resources, Conservation and Recycling*, 15:65–78.
198. WHISLER, W.D. (1967). A stochastic inventory model for rented equipment. *Management Science*, 13(9):640–647.
199. XEROX (1998). Annual corporate environmental report.
200. YUAN, X.-M. AND CHEUNG, K.L. (1998). Modeling returns of merchandise in an inventory system. *OR Spektrum*, 20(3):147–154.
201. ZEID, I., GUPTA, S.M., AND BARDASZ, T. (1999). A case-based reasoning approach to planning for disassembly. *International Journal of Intelligent Manufacturing*. (To appear).
202. ZHENG, Y.-S. (1991). A simple proof for optimality of (s, S) policies in infinite-horizon inventory systems. *Journal of Applied Probability*, 28:802–810.
203. ZHENG, Y.-S. AND FEDERGRUEN, A. (1991). Finding optimal (s, S) policies is about as simple as evaluating a single policy. *Operations Research*, 39(4):654–665.

Vol. 399: F. Gori, L. Geronazzo, M. Galeotti (Eds.), Nonlinear Dynamics in Economics and Social Sciences. Proceedings, 1991. VIII, 367 pages. 1993.

Vol. 400: H. Tanizaki, Nonlinear Filters. XII, 203 pages. 1993.

Vol. 401: K. Mosler, M. Scarsini, Stochastic Orders and Applications. V, 379 pages. 1993.

Vol. 402: A. van den Elzen, Adjustment Processes for Exchange Economies and Noncooperative Games. VII, 146 pages. 1993.

Vol. 403: G. Brennscheidt, Predictive Behavior. VI, 227 pages. 1993.

Vol. 404: Y.-J. Lai, Ch.-L. Hwang, Fuzzy Multiple Objective Decision Making. XIV, 475 pages. 1994.

Vol. 405: S. Komlósi, T. Rapcsák, S. Schaible (Eds.), Generalized Convexity. Proceedings, 1992. VIII, 404 pages. 1994.

Vol. 406: N. M. Hung, N. V. Quyen, Dynamic Timing Decisions Under Uncertainty. X, 194 pages. 1994.

Vol. 407: M. Ooms, Empirical Vector Autoregressive Modeling. XIII, 380 pages. 1994.

Vol. 408: K. Haase, Lotsizing and Scheduling for Production Planning. VIII, 118 pages. 1994.

Vol. 409: A. Sprecher, Resource-Constrained Project Scheduling. XII, 142 pages. 1994.

Vol. 410: R. Winkelmann, Count Data Models. XI, 213 pages. 1994.

Vol. 411: S. Dauzère-Péres, J.-B. Lasserre, An Integrated Approach in Production Planning and Scheduling. XVI, 137 pages. 1994.

Vol. 412: B. Kuon, Two-Person Bargaining Experiments with Incomplete Information. IX, 293 pages. 1994.

Vol. 413: R. Fiorito (Ed.), Inventory, Business Cycles and Monetary Transmission. VI, 287 pages. 1994.

Vol. 414: Y. Crama, A. Oerlemans, F. Spieksma, Production Planning in Automated Manufacturing. X, 210 pages. 1994.

Vol. 415: P. C. Nicola, Imperfect General Equilibrium. XI, 167 pages. 1994.

Vol. 416: H. S. J. Cesar, Control and Game Models of the Greenhouse Effect. XI, 225 pages. 1994.

Vol. 417: B. Ran, D. E. Boyce, Dynamic Urban Transportation Network Models. XV, 391 pages. 1994.

Vol. 418: P. Bogetoft, Non-Cooperative Planning Theory. XI, 309 pages. 1994.

Vol. 419: T. Maruyama, W. Takahashi (Eds.), Nonlinear and Convex Analysis in Economic Theory. VIII, 306 pages. 1995.

Vol. 420: M. Peeters, Time-To-Build. Interrelated Investment and Labour Demand Modelling. With Applications to Six OECD Countries. IX, 204 pages. 1995.

Vol. 421: C. Dang, Triangulations and Simplicial Methods. IX, 196 pages. 1995.

Vol. 422: D. S. Bridges, G. B. Mehta, Representations of Preference Orderings. X, 165 pages. 1995.

Vol. 423: K. Marti, P. Kall (Eds.), Stochastic Programming. Numerical Techniques and Engineering Applications. VIII, 351 pages. 1995.

Vol. 424: G. A. Heuer, U. Leopold-Wildburger, Silverman's Game. X, 283 pages. 1995.

Vol. 425: J. Kohlas, P.-A. Monney, A Mathematical Theory of Hints. XIII, 419 pages, 1995.

Vol. 426: B. Finkenstädt, Nonlinear Dynamics in Economics. IX, 156 pages. 1995.

Vol. 427: F. W. van Tongeren, Microsimulation Modelling of the Corporate Firm. XVII, 275 pages. 1995.

Vol. 428: A. A. Powell, Ch. W. Murphy, Inside a Modern Macroeconometric Model. XVIII, 424 pages. 1995.

Vol. 429: R. Durier, C. Michelot, Recent Developments in Optimization. VIII, 356 pages. 1995.

Vol. 430: J. R. Daduna, I. Branco, J. M. Pinto Paixão (Eds.), Computer-Aided Transit Scheduling. XIV, 374 pages. 1995.

Vol. 431: A. Aulin, Causal and Stochastic Elements in Business Cycles. XI, 116 pages. 1996.

Vol. 432: M. Tamiz (Ed.), Multi-Objective Programming and Goal Programming. VI, 359 pages. 1996.

Vol. 433: J. Menon, Exchange Rates and Prices. XIV, 313 pages. 1996.

Vol. 434: M. W. J. Blok, Dynamic Models of the Firm. VII, 193 pages. 1996.

Vol. 435: L. Chen, Interest Rate Dynamics, Derivatives Pricing, and Risk Management. XII, 149 pages. 1996.

Vol. 436: M. Klemisch-Ahlert, Bargaining in Economic and Ethical Environments. IX, 155 pages. 1996.

Vol. 437: C. Jordan, Batching and Scheduling. IX, 178 pages. 1996.

Vol. 438: A. Villar, General Equilibrium with Increasing Returns. XIII, 164 pages. 1996.

Vol. 439: M. Zenner, Learning to Become Rational. VII, 201 pages. 1996.

Vol. 440: W. Ryll, Litigation and Settlement in a Game with Incomplete Information. VIII, 174 pages. 1996.

Vol. 441: H. Dawid, Adaptive Learning by Genetic Algorithms. IX, 166 pages.1996.

Vol. 442: L. Corchón, Theories of Imperfectly Competitive Markets. XIII, 163 pages. 1996.

Vol. 443: G. Lang, On Overlapping Generations Models with Productive Capital. X, 98 pages. 1996.

Vol. 444: S. Jørgensen, G. Zaccour (Eds.), Dynamic Competitive Analysis in Marketing. X, 285 pages. 1996.

Vol. 445: A. H. Christer, S. Osaki, L. C. Thomas (Eds.), Stochastic Modelling in Innovative Manufactoring. X, 361 pages. 1997.

Vol. 446: G. Dhaene, Encompassing. X, 160 pages. 1997.

Vol. 447: A. Artale, Rings in Auctions. X, 172 pages. 1997.

Vol. 448: G. Fandel, T. Gal (Eds.), Multiple Criteria Decision Making. XII, 678 pages. 1997.

Vol. 449: F. Fang, M. Sanglier (Eds.), Complexity and Self-Organization in Social and Economic Systems. IX, 317 pages, 1997.

Vol. 450: P. M. Pardalos, D.W. Hearn, W.W. Hager, (Eds.), Network Optimization. VIII, 485 pages, 1997.

Vol. 451: M. Salge, Rational Bubbles. Theoretical Basis, Economic Relevance, and Empirical Evidence with a Special Emphasis on the German Stock Market.IX, 265 pages. 1997.

Vol. 452: P. Gritzmann, R. Horst, E. Sachs, R. Tichatschke (Eds.), Recent Advances in Optimization. VIII, 379 pages. 1997.

Vol. 453: A. S. Tangian, J. Gruber (Eds.), Constructing Scalar-Valued Objective Functions. VIII, 298 pages. 1997.

Vol. 454: H.-M. Krolzig, Markov-Switching Vector Autoregressions. XIV, 358 pages. 1997.

Vol. 455: R. Caballero, F. Ruiz, R. E. Steuer (Eds.), Advances in Multiple Objective and Goal Programming. VIII, 391 pages. 1997.

Vol. 456: R. Conte, R. Hegselmann, P. Terna (Eds.), Simulating Social Phenomena. VIII, 536 pages. 1997.

Vol. 457: C. Hsu, Volume and the Nonlinear Dynamics of Stock Returns. VIII, 133 pages. 1998.

Vol. 458: K. Marti, P. Kall (Eds.), Stochastic Programming Methods and Technical Applications. X, 437 pages. 1998.

Vol. 459: H. K. Ryu, D. J. Slottje, Measuring Trends in U.S. Income Inequality. XI, 195 pages. 1998.

Vol. 460: B. Fleischmann, J. A. E. E. van Nunen, M. G. Speranza, P. Stähly, Advances in Distribution Logistic. XI, 535 pages. 1998.

Vol. 461: U. Schmidt, Axiomatic Utility Theory under Risk. XV, 201 pages. 1998.

Vol. 462: L. von Auer, Dynamic Preferences, Choice Mechanisms, and Welfare. XII, 226 pages. 1998.

Vol. 463: G. Abraham-Frois (Ed.), Non-Linear Dynamics and Endogenous Cycles. VI, 204 pages. 1998.

Vol. 464: A. Aulin, The Impact of Science on Economic Growth and its Cycles. IX, 204 pages. 1998.

Vol. 465: T. J. Stewart, R. C. van den Honert (Eds.), Trends in Multicriteria Decision Making. X, 448 pages. 1998.

Vol. 466: A. Sadrieh, The Alternating Double Auction Market. VII, 350 pages. 1998.

Vol. 467: H. Hennig-Schmidt, Bargaining in a Video Experiment. Determinants of Boundedly Rational Behavior. XII, 221 pages. 1999.

Vol. 468: A. Ziegler, A Game Theory Analysis of Options. XIV, 145 pages. 1999.

Vol. 469: M. P. Vogel, Environmental Kuznets Curves. XIII, 197 pages. 1999.

Vol. 470: M. Ammann, Pricing Derivative Credit Risk. XII, 228 pages. 1999.

Vol. 471: N. H. M. Wilson (Ed.), Computer-Aided Transit Scheduling. XI, 444 pages. 1999.

Vol. 472: J.-R. Tyran, Money Illusion and Strategic Complementarity as Causes of Monetary Non-Neutrality. X, 228 pages. 1999.

Vol. 473: S. Helber, Performance Analysis of Flow Lines with Non-Linear Flow of Material. IX, 280 pages. 1999.

Vol. 474: U. Schwalbe, The Core of Economies with Asymmetric Information. IX, 141 pages. 1999.

Vol. 475: L. Kaas, Dynamic Macroeconomics with Imperfect Competition. XI, 155 pages. 1999.

Vol. 476: R. Demel, Fiscal Policy, Public Debt and the Term Structure of Interest Rates. X, 279 pages. 1999.

Vol. 477: M. Théra, R. Tichatschke (Eds.), Ill-posed Variational Problems and Regularization Techniques. VIII, 274 pages. 1999.

Vol. 478: S. Hartmann, Project Scheduling under Limited Resources. XII, 221 pages. 1999.

Vol. 479: L. v. Thadden, Money, Inflation, and Capital Formation. IX, 192 pages. 1999.

Vol. 480: M. Grazia Speranza, P. Stähly (Eds.), New Trends in Distribution Logistics. X, 336 pages. 1999.

Vol. 481: V. H. Nguyen, J. J. Strodiot, P. Tossings (Eds.). Optimation. IX, 498 pages. 2000.

Vol. 482: W. B. Zhang, A Theory of International Trade. XI, 192 pages. 2000.

Vol. 483: M. Königstein, Equity, Efficiency and Evolutionary Stability in Bargaining Games with Joint Production. XII, 197 pages. 2000.

Vol. 484: D. D. Gatti, M. Gallegati, A. Kirman, Interaction and Market Structure. VI, 298 pages. 2000.

Vol. 485: A. Garnaev, Search Games and Other Applications of Game Theory. VIII, 145 pages. 2000.

Vol. 486: M. Neugart, Nonlinear Labor Market Dynamics. X, 175 pages. 2000.

Vol. 487: Y. Y. Haimes, R. E. Steuer (Eds.), Research and Practice in Multiple Criteria Decision Making. XVII, 553 pages. 2000.

Vol. 488: B. Schmolck, Ommitted Variable Tests and Dynamic Specification. X, 144 pages. 2000.

Vol. 489: T. Steger, Transitional Dynamics and Economic Growth in Developing Countries. VIII, 151 pages. 2000.

Vol. 490: S. Minner, Strategic Safety Stocks in Supply Chains. XI, 214 pages. 2000.

Vol. 491: M. Ehrgott, Multicriteria Optimization. VIII, 242 pages. 2000.

Vol. 492: T. Phan Huy, Constraint Propagation in Flexible Manufacturing. IX, 258 pages. 2000.

Vol. 493: J. Zhu, Modular Pricing of Options. X, 170 pages. 2000.

Vol. 494: D. Franzen, Design of Master Agreements for OTC Derivatives. VIII, 175 pages. 2001.

Vol. 495: I Konnov, Combined Relaxation Methods for Variational Inequalities. XI, 181 pages. 2001.

Vol. 496: P. Weiß, Unemployment in Open Economies. XII, 226 pages. 2001.

Vol. 497: J. Inkmann, Conditional Moment Estimation of Nonlinear Equation Systems. VIII, 214 pages. 2001.

Vol. 498: M. Reutter, A Macroeconomic Model of West German Unemployment. X, 125 pages. 2001.

Vol. 499: A. Casajus, Focal Points in Framed Games. XI, 131 pages. 2001.

Vol. 500: F. Nardini, Technical Progress and Economic Growth. XVII, 191 pages. 2001.

Vol. 501: M. Fleischmann, Quantitative Models for Reverse Logistics. XI, 181 pages. 2001.